Y0-BVP-150

Technology Change and Female Labour in Japan

Technology Transfer, Transformation, and Development:
The Japanese Experience
Project Coordinator, Takeshi Hayashi

General Trading Companies: A Comparative and Historical Study, ed. Shin'ichi Yonekawa
Industrial Pollution in Japan, ed. Jun Ui
Irrigation in Development: The Social Structure of Water Utilization in Japan, ed. Akira Tamaki, Isao Hatate, and Naraomi Imamura
Technological Innovation and the Development of Transportation in Japan, ed. Hirofumi Yamamoto
The Japanese Experience in Technology: From Transfer to Self-reliance, Takeshi Hayashi
The Role of Labour-intensive Sectors in Japanese Industrialization, ed. Johzen Takeuchi
Technology Change and Female Labour in Japan, ed. Masanori Nakamura
Vocational Education in the Industrialization of Japan, ed. Toshio Toyoda

The United Nations University (UNU) is an organ of the United Nations established by the General Assembly in 1972 to be an international community of scholars engaged in research, advanced training, and the dissemination of knowledge related to the pressing global problems of human survival, development, and welfare. Its activities focus mainly on peace and conflict resolution, development in a changing world, and science and technology in relation to human welfare. The University operates through a worldwide network of research and postgraduate training centres, with its planning and coordinating headquarters in Tokyo, Japan.

The United Nations University Press, the publishing division of the UNU, publishes scholarly books and periodicals in the social sciences, humanities, and pure and applied natural sciences related to the University's research.

Technology Change and Female Labour in Japan

Edited by
Masanori Nakamura

TOKYO • NEW YORK • PARIS

The United Nations University project on Technology Transfer, Transformation, and Development: The Japanese Experience was carried out from 1978 to 1982. Its objective was to contribute to an understanding of the process of technological development in Japan as a case-study. The project enquired into the infrastructure of technology, human resources development, and social and economic conditions and analysed the problems of technology transfer, transformation, and development from the time of the Meiji Restoration to the present. The research was undertaken by more than 120 Japanese specialists and covered a wide range of subjects, including iron and steel, transportation, textiles, mining, financial institutions, rural and urban society, small industry, the female labour force, education, and technology policy.

This volume examines the link between technological innovation and female labour from the standpoint of the relationship between modes of production and forms of labour.

The United Nations University gratefully acknowledges the generous support of the Japan Foundation for the United Nations University in the publication of this book.

This book was originally published in Japanese in 1985 by the United Nations University under the title *Gijutsu kakushin to joshi rōdō*.

United Nations University Press
The United Nations University, 53-70, Jingumae 5-chome,
Shibuya-ku, Tokyo 150, Japan
Tel: (03)3499-2811 Fax: (03)3499-2828
Telex: J25442 Cable: UNATUNIV TOKYO

Typeset by Asco Trade Typesetting Limited, Hong Kong
Printed by Permanent Typesetting and Printing Co., Ltd., Hong Kong

UNUP-553
ISBN 92-808-0553-3
United Nations Sales No. E.93.III.A.7
03300 C

Contents

Foreword

This book represents part of the results of research carried out by the Institute of Developing Economies at the request of the United Nations University for the Project on Technology Transfer, Transformation and Development: The Japanese Experience.

Anyone with an interest in development issues, and anyone who has ever been to a developing country, is aware of the importance of human resources, and development experts have been talking about the issue for many years. The aim of this project is to examine, from the general viewpoint of development, Japan's experience in undergoing an industrial revolution with the help of technology transferred from the advanced countries. It was natural that the focus should be on the labour force. However, we had a specific reason for addressing the subject of female labour rather than limiting the scope of our study to the general discussion of labour-force creation and skill development. In the words of Shōji Okumura, a historian of technology, it is historically undeniable that the dynamism of the present-day Japanese economy began with the energy of the nimble fingers of its young women in the latter decades of the nineteenth century.

Further, although new technology did in some cases occasion loss of work opportunities for women, technological innovation and the female employment rate generally move along parallel lines. In fact, when advanced technology is adopted, women often become the core labour force. Re-entry into the labour force after taking time out for marriage and child-rearing has become an established social phenomenon over the past 30 years or so. It should be noted here that Japanese statistics treat the concept of "unemployment" differently from other countries: people seeking employment for the first time are not counted among the unemployed, and this shrinks the overall figures for the unemployed. The increase in the number of female job-seekers reflects changes in their life cycle and lifestyles. There are now (in 1990) more than 15 million working housewives, a number far exceeding that of housewives occupied exclusively at home, and the unmarried female workforce is a minority in the labour market.

The culture of some countries rejects female employment, particularly as paid labour. This pattern was found in Japan as well before the Second World War, but the need to support a rapidly growing population made it imperative to adopt more advanced technology, and changes in technology promoted the employment of women. This inevitably produced shifts in the basic social structure of traditional culture, and such changes will undoubtedly continue. Japan was unable to maintain its indigenous culture untouched while pursuing industrial development and technological innovation. Other nations' cultures may be able to do so, but the prospect is not optimistic. The modernization of technology does not have to mean the complete reorganization of traditional culture. It can promote the partial advancement and more widespread transmission of culture in some respects, in some cases reinvigorating it in a more sophisticated form.

The Japanese did not cling to old ways. The national consensus welcoming the transfer of modern technology was in place more than a hundred years ago, although this was undoubtedly because people were already relatively poor and subject to population pressures, and because they recognized that this was the only alternative to ensure survival in the then international setting. Some scholars say that an "industrious revolution" took place prior to the industrial revolution, and that women, and even children, were the indispensable core of labour. This state of affairs continued in the course of the industrial revolution, as the earnings of the household head were not enough to support the family, and women and children also had to work. Some maintain that this "employment of entire families" was in complete opposition to the concept of full employment, but it was a common pattern before the war in small and medium-size factories, and in those smaller workplaces where Japan's industrial technology (and skills) developed.

The situation today has changed completely. Women are no longer engaged in hard labour, and office automation and micro-electronics are promoting a rapid influx of women into clerical or service sectors. But the foundations for these trends were laid by the increase in the nuclear family, rapid urbanization, and higher education, and further facilitated by the greatly reduced need for household labour as a result of the developments in the food and apparel industries. On the other hand, there is also a marked trend towards the part-time or indirect employment of women, and they are playing the role of buffer during economic fluctuations. This will surely necessitate a revision of non-Japanese ideas about "Japanese-style management," in particular, the system of lifetime employment. With the exception of professional and career workers, women are usually not employed for life.

The development of mechanization does not lead automatically to the lightening of the burden of female labour. In Japan, the mechanization of farming is encouraging men to take jobs outside agriculture and making it a predominantly female occupation. In fishing villages, the motorization of fishing boats has led to wives taking their place as indispensable working partners of their husbands. Among the trade-offs are new health problems among women young and old in farming and fishing villages. In the past, young women endured long workdays (the 12-hour day was standard) in silk-reeling or textile plants, ruining their health for the sake of meagre earnings and poor nutrition which helped to secure the survival of their parents and siblings in poor fishing or farming villages. Today, although the nature and status of problems have changed, they remain in a sophisticated modern form at a higher level and are perpetuated.

Some of the young women who, by virtue of their robust health, survived the hardships of the silk-reeling industry around the turn of the century were still alive when our project was initiated. Among their recollections is *Aa, nomugi tōge* [Remember the Nomugi Pass] (1977) by Shigemi Yamamoto, considered the finest example of documentary literature. This record of the not-too-distant past moved many after it had been made into a film and shown in other countries.

This book also touches on the same period, and other groups in this research project are dealing with it. For example, Takeo Izumi's excellent work on (female) labour in the textile industry has been translated (*The Developing Economies*, IDE, Tokyo, 1979) and I recommend that it be read along with this book. This project series also includes a volume dealing with the issue of educational systems and human resources.

As Professor Masanori Nakamura, the editor of *Technology Change and Female Labour in Japan*, points out, this book's approach is unique for Japan, and few other books have covered the subject from the start of industrialization to the present day. I am grateful for the pioneering spirit of the editor and of the authors of each chapter. I also sincerely hope that research on the issues raised will be undertaken more extensively and in further depth and detail.

This book would not have been possible without the dedicated efforts of my colleagues, in particular Akiko Akemine, who bore the brunt of the actual editing work. I would also like to thank Takeo Uchida and Shigeo Minowa of the United Nations University for their support and cooperation.

Takeshi Hayashi

Preface

Female labour has become an indispensable element of today's industrial society. During the 1960s, as Japan's rapid economic growth rose to its peak, the entry of women into the workforce in large numbers brought about major changes in the country's employment structure. The high tempo of innovation in science and technology, needless to say, formed the backdrop for this trend; the spread of office automation and use of electronic equipment expanded employment opportunities for women and accelerated changes in the woman's life cycle. Technological development transformed the production process in factories and labour–employer relations; it led even to transformations in human relations in society and in the household environment. The Japanese have never before experienced a technological and social revolution of this degree, occurring this rapidly. In this context, it was only natural that "Technology Change and Female Labour in Japan" should be added to the list of topics to be studied in the United Nations University Project on Technology Transfer, Transformation, and Development: The Japanese Experience" in cooperation with the Institute for Developing Economies.

IDE's project coordinator, Professor Takeshi Hayashi, approached me in the spring of 1981 to request me to undertake a collaborative study on technological change and female labour. I was finally convinced to undertake the project because of the topicality as well as the great importance of the subject, and work began in March that year. I and the members of our group (the other scholars collaborating on the study) held regular seminars every two or three months to exchange ideas. It seemed to us that studies showing how technology had changed in each period and in specific industries, and how those changes had affected the nature of female labour in Japan, would offer a useful guide to those considering the problems of female labour in developing countries currently in the throes of rapid technological change. We discussed this challenge many times in the room set aside for our meetings at the Institute. About one year after these meetings began, we decided on our respective topics for writing. The draft manuscripts were prepared by the end of 1983 and, after being read by the editors, were returned to the authors for checking and revision.

The study had to proceed in accordance with four important considerations: (1) our small study group had to cover the period from the beginning of Japan's modernization (Meiji, 1868–1912) to the rapid economic growth period (1955–1973); (2) there was surprisingly little previous research on the topic; (3) reliable documents and statistics that could be used in the study were quite limited; and (4) the resulting research should be presented in such a way as to be useful to developing countries. With these concerns in mind, we determined to define the focus of the study very specifically; for the pre-war period, we selected the three topics of silk-reeling labour, women workers in the coal industry, and female workers of the urban lower class. For the post-war period, we decided to deal with women working in family-based industries, as exemplified by agriculture and fisheries, and with female employed labour during the rapid economic growth period. This book is made up of five chapters, each covering one of these topics, along with an introductory chapter providing an overview of female labour in the pre-second World War period, and a concluding chapter. The following is a brief outline of the book's content.

The introductory chapter deals with female labour in the industrial revolution period from the two perspectives of production type and source of labour supply, divided into six categories, and the changes for each during the First World War period, the inter-war period, and the wartime (China War and Pacific War) period. Chapter 1 focuses on the conditions of workers in the silk-reeling industry, reflecting developments in production technology from the end of the transition period to the modern Meiji period government in around 1900, mainly as observed in the Suwa region of Nagano Prefecture. Chapter 2 describes female workers in the coal-mining industry, explaining how technological innovation prescribed labour management in the coal-mining industry from 1900 to the 1930s and analysing the impact of those changes on female labour. Chapter 3 looks at female labour among the lower classes in the large cities from the 1870s to the 1920s, examining how industrialization and the progress of urbanization altered the framework of daily life and the structure of employment for these women. The next two chapters cover the post-war period. Chapter 4 deals with family-based enterprises in agriculture and fisheries, portraying the impact of technological innovation during the rapid economic growth period on employment patterns among women of agricultural, and especially fisheries, households. Chapter 5 discusses the sharp rise in female employment during the rapid economic growth period and the changes in Japanese lifestyles and living patterns brought about by the automation of labour processes and technological innovation.

The concluding chapter summarizes the findings of each of the studies and places them in the context of historical research, and examines their significance and usefulness from the point of view of the problems of female labour in other Asian countries.

As reflected in the above remarks, female labour encompasses both employed and family-based workers, but, with the exception of chapter 4, which examines women working in agriculture and fisheries in the post-war period, the main thrust of analysis is on employed workers. This book was originally compiled for an overseas readership, mainly for the reference of third-world countries, and plans did not include the publication of Japanese or English editions. As a result, the authors were required to include references and information which are often common knowledge to Japanese specialists. They therefore made full use of existing studies, including quotations and citations from these works.

The fact that the study of how technology transfer, transformation, and development changed the nature of female labour was a completely new field left us relatively unconstrained by previous research and gave us a free hand in our endeavours. We will be satisfied if this work provides a useful landmark in further efforts to open up this field of research.

Masanori Nakamura

Introduction: Types of Female Labour and Changes in the Workforce, 1890–1945

Yutaka Nishinarita

This book covers two periods of modern Japanese history: from the Meiji Restoration of 1868 to the end of the Second World War, and from 1945 to 1985. For women workers after 1945, the studies (chapters 1–3) present a relatively clear picture of the overall situation. For the pre-war period, however, the essays (chapters 4 and 5) focus on female workers in specific industries—silk-reeling and coal-mining—and on the occupations of the urban lower class, and do not attempt a comprehensive treatment.

Before 1945, the majority of women worked in the agricultural and extractive industries (fig. 1). The number of women in the primary sector dropped from 7 million (more than 70 per cent of all women workers) before the First World War to 6 million during the war years. After 1923, when it was at its lowest, however, their number increased steadily, reaching about 60 per cent of all women workers in the 1930s. Female labour in the tertiary (service) sector, especially sales, shows a steady increase, expanding from 8–9 per cent of the total female workforce before the First World War to 14–15 per cent after the war. Immediately after the financial crisis of 1927, the tertiary sector employed 18 per cent of all women workers, and in 1928 the figure exceeded that in manufacturing.

The number of women working in the industrial sector grew during the First World War and in the late 1930s, but declined or levelled off during the intervening period, remaining at around 15 per cent. If we define female labour broadly to include women working in family enterprises, before 1945 the vast majority were engaged in family enterprises, including farming and small business, in the primary and tertiary sectors. Only a very small number worked as hired labour in the manufacturing sector.

In the pre-war period, female hired labour bore the brunt of technological change. Only after 1945 were women in traditional family enterprises affected by technological innovation (see chapter 4). Here, the discussion will be limited primarily to female wage-labour. Three general methodological observations may be made.

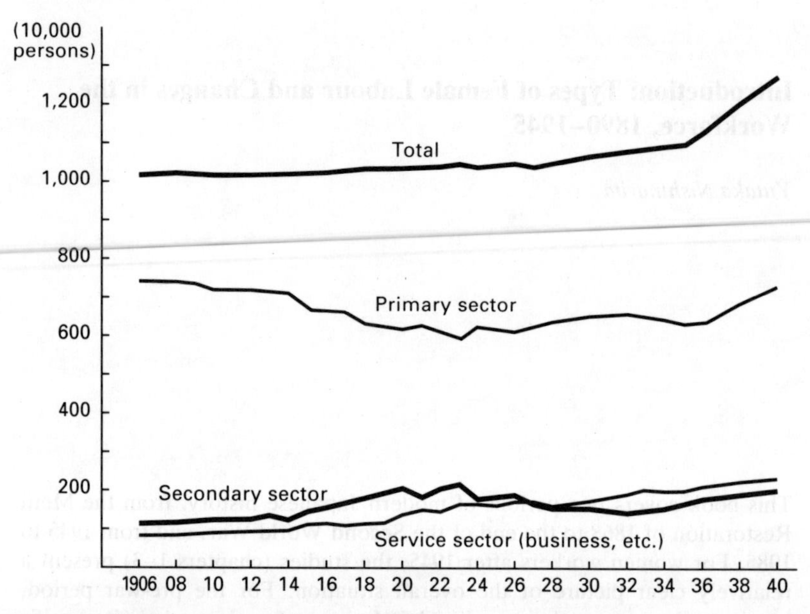

Fig. 1. Number of Women Workers by Industrial Sector, 1906–1940
Notes: Figures are estimations.
Source: Compiled from tables provided in Umemura Mataji, "Sangyōbetsu koyō no hendō: 1880–1940 nen" [Trends in Industrial Sector Employment: 1880 to 1940], *Keizai kenkyū*, vol. 24, no. 2 (1973).

First, female workers may be categorized in a number of ways, depending on one's theoretical and methodological assumptions. Here, I will classify women workers according to the form of production (labour and technology) and social supply of labour (labour market).

Second, I will look at the transformation, and in some cases dissolution, of each category of women workers in the course of the evolution of Japanese capitalism, attempting to identify the structural dynamics involved. To clarify this process, I divide the development of the female labour force into four periods: the industrial revolution (1880s–1907), the First World War (1914–1918), the inter-war period (1920–1936), and the Second Sino-Japanese and Pacific War period (1937–1945).

Finally, looking at female labour after 1945 we can obtain a comprehensive grasp of the continuities and discontinuities between the pre-war and post-war periods. As this paper focuses on the pre-war situation, the subject of continuity-discontinuity will be briefly touched on in the conclusion of this introductory chapter, providing a link to chapter 5.

I. The Industrial Revolution

Japan's industrial revolution took place in the 20 years before about 1907. The revolution was led by the cotton and silk industries. Cotton-spinning relied on imported cotton and spinning machines, and from the start production was large-scale and mechanized. The industry grew rapidly, displacing both hand-spun and imported cotton yarn. In 1897, yarn exports exceeded imports for the first time, indicating that capitalist production had transformed cotton-spinning into an export industry.

Silk-reeling, too, was mechanized but developed from a combination of Western and indigenous technology. Selling primarily to the North American market, silk yarn became Japan's main earner of foreign reserve revenue. Between 1906 and 1910, Japanese silk exports overtook even those of China, until then the world's largest silk-yarn exporter.

In contrast to cotton-spinning and silk-reeling, civilian machinery and equipment manufacturing remained in an incipient stage of development during the years of early industrialization. Machinery manufacturing acquired an abnormal structure heavily skewed by a surge forward in military-related development. Military procurement for the army and navy during the Russo-Japanese War (1904–1905), however, proved an important stimulus to civilian machine manufacture, and shipbuilding in particular advanced rapidly. By 1907, with the completion of the *Ten'yō maru*, Japan's first modern domestic-built steam vessel, the industry had risen to contemporary world standards. As Japan's capacity for self-sufficiency in shipbuilding increased, a number of large machine manufacturers emerged, and Japanese craftsmen began producing American-style machine lathes on their own.

By around 1907, the way seemed clear for the establishment of the machine and equipment industry, and the foundations for development of an independent national economy, based on the internal linkage between the consumer goods and industrial goods industries, took shape. In fact, however, this linkage remained extremely weak, and in the cotton and silk industry, for example, machine manufacturers could not satisfy domestic demand for spinning and reeling equipment until the 1930s. Industrial capitalism depended on imports for many heavy industrial goods, and these were paid for largely out of the earnings from raw-silk thread exports. The reeling industry, then, financed the machine imports essential for the development and expansion of industrial production in Japan.

These features of incipient Japanese capitalism are illustrated in the industrial structure. By 1909, the end of the industrial revolution, the spinning and weaving industry, including both cotton-spinning and silk-reeling, accounted for 51 per cent of the value of total industrial production and employed 64 per cent of the workforce (table 1). The metalworking and machine industries produced only 10 per cent of that value and hired a mere 8 per cent of the labour force. The industrial structure was lopsided, skewed heavily toward spinning and weaving.

Table 1. Amount of Production and Number of Workers by Industry, 1909–1940

	1909	1914	1919	1925	1929	1935	1940
Amount of production (1,000 yen)							
Spinning and weaving industries							
Silk-spinning	111,561 (14.0)[b]	166,438 (21.1)	846,527 (12.3)	937,139 (13.5)	864,353 (11.2)	484,587 (4.5)	956,383 (3.3)
Cotton-spinning	121,218 (15.2)	203,722 (14.8)	760,476 (11.0)	781,442 (11.3)	822,035 (10.6)	877,043 (8.1)	870,126 (3.0)
Silk textiles	46,234 (5.8)	55,151 (4.0)	453,122 (6.6)	275,324 (4.0)	248,503 (3.2)	237,513 (2.2)	606,062 (2.1)
Cotton textiles	61,975 (7.8)	117,485 (8.6)	795,350 (11.5)	709,211 (10.2)	485,393 (6.3)	641,148 (5.9)	388,214 (1.3)
Other textiles	403,452 (50.7)	660,175 (48.1)	3,514,386 (51.0)	3,479,416 (50.2)	3,323,137 (42.9)	3,497,652 (32.3)	4,976,151 (17.1)
Metalworking and machines/appliances	76,249 (9.6)	184,681 (13.5)	1,274,069 (18.5)	934,994 (13.5)	1,450,231 (18.7)	3,350,433 (31.0)	14,343,557 (49.3)
Chemicals	80,172 (10.1)	164,125 (12.0)	723,493 (10.5)	771,593 (11.1)	1,044,266 (13.5)	1,814,724 (16.8)	4,623,270 (15.9)
Processed food	147,240 (18.5)	221,246 (16.1)	742,997 (10.8)	1,102,313 (15.9)	1,163,314 (15.0)	1,168,479 (10.8)	2,465,196 (8.5)
Misc. manufacturing[c]	89,316 (11.2)	142,201 (10.4)	634,465 (9.2)	636,595 (9.2)	757,772 (9.8)	984,529 (9.1)	2,684,528 (9.2)
Total	796,429 (100.0)	1,372,429 (100.0)	6,889,410 (100.0)	6,924,911 (100.0)	7,738,720 (100.0)	10,815,817 (100.0)	29,092,702 (100.0)
Number of workers							
Spinning and weaving industries							
Silk-reeling	191,561 (24.5)	224,287 (23.8)	297,957 (18.6)	343,654 (19.1)	416,715 (22.9)	280,508 (11.9)	198,837 (5.2)
Cotton-spinning	89,781 (11.5)	112,858 (12.0)	187,707 (11.7)	210,997 (11.7)	179,558 (9.9)	168,800 (7.1)	138,203 (3.6)
Silk textiles	59,574 (7.6)	45,649 (4.8)	93,453 (5.8)	74,611 (4.1)	67,048 (3.7)	86,385 (3.7)	125,955 (3.3)
Cotton textiles	71,759 (9.2)	88,662 (9.4)	160,426 (10.0)	168,016 (9.3)	126,495 (7.0)	139,128 (5.9)	75,769 (2.0)
Other textiles	501,538 (64.3)	583,469 (62.0)	917,238 (57.1)	1,004,317 (55.7)	1,037,829 (57.1)	1,071,188 (45.4)	998,217 (26.1)
Machines and							1,726,123 (45.1)
appliances	65,017 (8.3)	102,257 (10.9)	273,899 (17.1)	317,306 (17.6)	281,033 (15.5)	583,833 (24.7)	385,264 (10.1)
Chemicals	27,399 (5.3)	40,212 (4.3)	98,449 (6.1)	98,084 (5.4)	122,330 (6.7)	226,960 (9.6)	222,483 (5.8)
Processed food	71,313 (9.1)	76,856 (8.2)	104,772 (6.5)	170,648 (9.5)	142,998 (7.9)	158,125 (6.7)	497,751 (13.0)
Misc. manufacturing[c]	115,227 (14.8)	138,808 (14.7)	211,384 (13.2)	211,651 (11.7)	232,794 (12.8)	320,781 (13.6)	3,829,835 (100.0)
Total	780,494 (100.0)	941,602 (100.0)	1,605,742 (100.0)	1,802,006 (100.0)	1,816,984 (100.0)	2,360,887 (100.0)	

a. Based on companies with five or more workers.
b. Numbers in parentheses are percentages.
c. "Misc. manufacturing" includes lumber and wood-processing, printing and bookbinding, ceramics, quarrying, and others.
Source: Compiled on the basis of data contained in documents in *Kōgyō tōkei go-jūnenshi, Shiryōhen 1* [Fifty Years of Industrial Statistics, Documents, vol. 1].

Table 2. Composition of the Female Labour Force by Industry

	1902	1907	1914	1919	1925	1929	1935	1940
Spinning and weaving industries	61,980 (79.4)[a]	68,273 (79.3)	100,460 (80.6)	175,873 (77.4)	199,372 (78.5)	185,280 (77.5)	205,725 (85.7)	188,783 (85.2)
Cotton-spinning	120,980 (93.8)	148,588 (94.9)	209,703 (95.0)	278,249 (93.4)	315,870 (91.5)	385,167 (92.4)	255,066 (92.0)	174,441 (92.2)
Silk-reeling	51,187 (86.6)	93,749 (87.2)	119,850 (85.2)	237,986 (81.6)	236,315 (81.2)	203,613 (81.2)	291,614 (82.3)	271,870 (83.2)
Weaving (A)[b]	730,213 (94.5)	726,232 (95.7)	575,797 (94.3)	951,834 (90.8)	539,015 (86.5)	454,467 (85.1)	492,777 (83.9)	?
Weaving (B)[b]	2,310 (75.3)	3,204 (71.7)	34,835 (69.2)	41,205 (52.9)	40,042 (47.8)	40,441 (44.4)	60,432 (44.7)	139,642 (58.9)
Other	983 (2.9)	2,395 (3.9)	4,184 (4.8)	16,561 (6.4)	19,623 (6.2)	19,995 (7.1)	48,219 (8.2)	175,293 (10.2)
Metalworking/machines								
Chemicals	43,683 (53.1)	22,386 (34.2)	28,101 (33.4)	56,248 (29.7)	48,464 (28.6)	53,348 (27.7)	98,792 (30.7)	126,448 (31.7)
Processed food	13,316 (44.2)	19,643 (40.8)	10,882 (18.6)	19,726 (18.8)	47,190 (27.7)	21,317 (14.9)	31,671 (20.0)	69,267 (31.1)
Misc. manufacturing	11,579 (35.8)	20,342 (38.4)	27,023 (34.3)	43,743 (30.6)	48,872 (28.4)	60,590 (29.9)	90,143 (31.1)	152,242 (30.0)
Total (including misc.) (excludes weaving (B))	313,269 (62.8)	385,936 (60.0)	535,297 (62.7)	870,797 (54.0)	955,827 (52.9)	969,835 (53.1)	1,081,702 (45.7)	1,298,059 (33.8)
Mining			67,291 (22.9)	111,849 (24.0)	72,321 (23.3)	55,104 (19.2)	25,389 (9.9)	55,240 (11.1)
Metals	?	?	14,893 (15.7)	15,167 (15.0)	5,895 (13.1)	5,077 (10.6)	6,433 (9.3)	18,358 (11.9)
Coal	?	?	51,400 (27.5)	95,283 (27.4)	65,402 (25.9)	49,277 (21.5)	17,847 (10.2)	34,431 (10.6)

a. Figures in parentheses show percentage of women workers *vis-à-vis* the entire labour force for that industry.

b. "Weaving (A)" means factory workers; "Weaving (B)" non-factory workers.

Source: Compiled from *Nōshōmu tokeihyō* [Statistics on Agriculture and Commerce Affairs], *Kōjō tōkeihyō* [Factory Statistics], *Shōkōshō tōkeihyō* [Ministry of Commerce and Industry Statistics], and *Honpō kōgyō sūsei* [Trends in Japan's Mining Industry].

The overwhelming majority of workers in the spinning and weaving industry were women, and most women in the labour force were employed in this sector. During the industrial revolution (surveyed in 1902 and 1907), female workers accounted for between 94 and 95 per cent of those employed in silk filatures, 79 per cent of the workers in cotton mills, and 87 per cent of factory weavers (table 2). A closer examination of the categories in table 2 shows the composition of the female working population to be more complex than it first appears. The extremely large number of women working in domestic manufactures ("Textiles (B)")—roughly 700,000—worked for putting-out operations controlled by *toiya* agents. The chemical industry hired between 20,000 and 40,000 women, and processed food and "miscellaneous manufacturing" each employed between 10,000 and 20,000. Women accounted for between 35 and 40 per cent of the workers in each of these industries.

In the chemical industry, women workers concentrated in the match factories. In the manufacture of food, beverages, and tobacco, most worked in tobacco factories, and in "miscellaneous manufacturing" they engaged mainly in straw-plaiting and the weaving of figured rush mats (*hanagoza*). According to a 1902 survey by the Ministry of Agriculture and Commerce (Nōshōmushō), women comprised 77 per cent of workers in the match industry and 86 per cent of those in the tobacco industry.[1] Although there are no confirming statistics, we may suppose that the coal industry, too, employed large numbers of women during the period of the industrial revolution.

Most of the women engaged in these activities (coal-mining was an exception) were under 20 years of age (table 3). With the exception of tobacco, all

Table 3. Age Composition of the Female Workforce, 1901 (percentages)

	Under 14	15–19	Over 20
Cotton-spinning[a]	11.4	41.6	47.0
Silk-reeling[b]	18.3	47.9	33.8
Textiles[c]	16.7	39.9	43.4
Matches[d]	18.6	40.3	41.1
Tobacco[e]	6.7	43.5	49.7

a. 19,344 workers in 16 factories in the Kansai area.
b. 12,519 workers in 205 factories in Nagano Prefecture.
c. 63,701 workers in weaving factories in Hachioji (Tokyo), Tango (Kyoto Prefecture), Sakai (*dantsū*) (Osaka), Ashikaga (Tochigi Prefecture), Nakajima (Aichi Prefecture), Fukui Prefecture (silk-weaving), and Fukuoka Prefecture (*kurume gasuri*).
d. 3,996 workers in 14 factories in Osaka.
e. 4,958 workers in 10 factories.
Source: Nōshōmushō Shōkō Kyoku Hen [Commerce and Industry Bureau, Ministry of Agriculture and Commerce], ed., *Shokkō jijō* [Conditions in the Textile Industry] (1903).

of the goods produced by these industries were for export.[2] Young female labour, concentrated in the leading export industries, formed a crucial link in the reproductive cycle of early industrial capitalism in Japan. Below, we shall summarize the main features of the female labour force in each sector, focusing on labour and technology (forms of production) and the social supply of labour (the labour market).

1. Cotton-spinning

The majority of female cotton-mill operatives were unmarried and less than 20 years old (table 3). In 1901, 59 per cent of the workers in England's spinning industry were women, and 60 per cent of these were over 20—and many of them were married.[3] The unusually high proportion of unwed adolescent women in the workforce is one of the distinguishing features of Japan's early cotton industry. This branch was able to absorb so many young women because of the introduction and rapid spread in the late 1880s of ring spindles, which were easier to operate than the more exacting mule spindles. Japan's quick switch to the technically advanced ring spindle gave it a technological edge over England, which relied exclusively on the mule, and this difference probably explains the larger proportion of older women in English cotton mills.

In the early stages of the industrial revolution, impoverished urban families and poor peasant producers on the periphery of urban centres provided cotton mills with female labour. But as the demand for labour expanded, factories cast recruitment nets wider, and by the late 1890s most operators were the daughters of poor peasants and tenant farmers who came to the cities from the rural hinterlands to find seasonal work (*dekasegi*).[4] The labour contracts these young women signed with the factories that hired them overwhelmingly favoured management. The "agreements" prohibited the worker from leaving before the specified period of employment was up and allowed the company to fine violators. Management, however, was free to dismiss workers as it pleased.[5] Such arbitrariness was determined by the fact that recruits, who were poor, were advanced a sum to cover travel and outfit expenses at the time of employment. This money had to be repaid out of wages and effectively bound the worker to the factory. Women in the cotton industry were in effect bonded workers, and as a result they were not independent sellers in the labour market. Under this system of employment, female mill operatives were forced to work a double shift extending far into the night, and for lower wages than their counterparts in colonial India.[6]

2. Silk-reeling

The silk filature industry developed rapidly on a traditional foundation to which foreign technical implants were added. Silk-reeling machinery combined the best features of Western and indigenous technology. But reeling machines remained sophisticated tools dependent, as in the past, on the nimble fingers and dexterity of their female operatives. For this reason, small-scale factories hiring between 20 and 40 women for skilled manual work formed the core of silk-reeling operations.

Silk-reelers, like cotton-spinners, were the daughters of poor peasants who emigrated to the cities to take seasonal jobs. As industrialization proceeded, they came from remote areas throughout Japan.[7] Unilateral labour contracts and the practice of advancing travel funds indentured these workers, like their counterparts in the cotton mills, to the filature factories. Operatives received low wages based on a specially graded wage system and were obliged to work long hours, an average of 13 to 14 hours a day, and sometimes as many as 17 and 18 hours.

3. Textiles

Although in some areas weaving had progressed beyond handicraft factory production, the putting-out system was prevalent: a large merchant house or enterprising peasant family, the *orimoto*, advanced yarn and money and loaned looms to smaller domestic producers, who were paid piece-rates for their output. Most weavers were farm women who wove cloth at home in their spare time or during the slack season to supplement income from agriculture.

A 1903 Ministry of Agriculture and Commerce report noted that: "Local women prefer to remain at home and work for piece-rates than to be bound to a factory for long periods of time. Moreover, female factory workers are looked down on."[8] For this reason, textile manufacturers found it difficult to recruit from nearby areas and were forced to recruit from more distant regions.[9] It took women factory workers between four and five years of training to learn how to operate a mechanical hand loom properly, so most enterprises had an apprenticeship system.

"An apprentice contract is made under the name of trainee weavers. Even when these trainees become regular employees no fixed wage is stipulated. The contract stated that an apprentice allowance would be paid only after the term of apprenticeship expired."[10] Contracts were extremely unilateral, severely limiting the freedom of weavers, as with labour contracts in the spinning and silk-reeling industries.[11]

4. Match Manufacturing

Most of the match production process was handled by female manual factory labour, though, later, matchstick alignment came to be done by indigenously developed machines. Some work, including matchbox-making, was left to people working at home through the medium of *toiya* agents. Virtually all match factory employees and home workers were the daughters of poor urban families living near the factories.[12]

5. Tobacco Manufacture

Female workers in the tobacco industry were engaged mainly in sorting and rolling tobacco leaves for cigars, rolling and wrapping leaves for cigarettes, and packaging. In the course of the industrial revolution, the work of cigarette-rolling and wrapping was gradually mechanized, as it was recorded that "[the work] once belonged solely to the realm of manual labour, but now that machines are used, it is the work of many female factory workers."[13] Mechanization not only changed the content of female labour but also brought about a decline in *toiya*-mediated tobacco-rolling on a piece basis, done by women at home.

> Piece-basis rolling at home was once a widespread practice. There was even a certain firm that set up branches and agents' offices at many places in the neighbouring prefectures to organize large amounts of hand-rolled tobacco done at home. As more of this work is now being done in factories, hand-rolling as a cottage industry is on the ebb. The introduction of machines, which makes hand-rolling unnecessary, has produced intense competition among tobacco-makers, who cannot survive without using machines.[14]

Generally, female workers in tobacco manufacturing commuted to nearby factories located in the cities. As indicated by a record which states that "their character is nobler than that of match-manufacturing workers",[15] the tobacco workers presumably came from a stratum of the poor (then called *saimin*) one rank higher than the poor of the urban slums.

6. Straw Mats and Other Products

Straw-plaiting (*sanada*) and weaving of figured rush mats (*hana-mushiro*) were done by farmers at their homes as a side job, under the control of *toiya* agents. There were signs for a while of emerging factory production in this realm too, but ultimately the agent-mediated home manufacturing prevailed. "The rise or fall of the *sanada* and *hana-mushiro* industry is dependent upon vigorous or inactive foreign trade. Partly for lack of foreign-trade know-how and partly because of a shortage of capital it often happens that the scale of business has to be either drastically expanded or drastically reduced. An incorporated enterprise system would not be flexible enough for that drastic adaptation."[16] The insecurity of relying on foreign trade was the condition that allowed *toiya* home manufacturing to survive.

7. Coal-mining

In coal-mining during the industrial revolution period, most mining processes continued to rely on manual labour using simple tools, although hoisting whims were in use for the main shafts. A working pattern in which a married couple worked as a unit, with the husband (*sakiyama*) digging out the ore and the wife (*atoyama*) assisting him by carrying away the coal, became widespread. Unlike in the industries discussed thus far, married women made up most of the female workforce in the coal-mining industry. Many left their native villages with their whole families to live near the mines where they worked. The *naya* or "stable" system emerged to supervise these miner families. This was an indirect labour-management system with the *naya* chief as an agent, who recruited miners, had them live in the bunk-houses he provided, and supervised their daily lives. Money borrowed in various forms from the chief kept the miners in bondage to him.

The above outlines the main features of female labour in seven industries. We can now categorize female workers during the industrial revolution period in terms of technology/labour (production form) and source of worker supply (labour market) into six types as follows:

- Type 1: Imported modern industry (large, mechanized factories) and seasonal work (cotton-spinning).
- Type 2: Manual factory manufacturing and seasonal work (silk-reeling and part of the textile industry).
- Type 3: *Toiya*-mediated home labour in rural areas (weaving, straw-plaiting, and rush-mat making).
- Type 4: Mechanized industry and urban poor (*saimin*) workers (tobacco manufacturing).
- Type 5: Manual factory manufacturing, *toiya*-mediated home labour, and urban-slum (*hinmin*) workers (match manufacturing).
- Type 6: Hard labour by families who had left their villages (coal-mining).

Overall, the female workers of the industrial revolution period lacked the independence of modern wage-earners, and displayed the features of pre-modern, agrarian-style, poor urban workers.

II. First World War Period

The First World War gave an unprecedented boost to Japanese capitalism. With the withdrawal of European countries from the Asian market and with the boom in the United States, Japanese exports of cotton goods and silk expanded, while the decline in import pressures led to remarkable progress in heavy and chemical industries. The rate of self-sufficiency in machines and appliances rose from 62 per cent in 1909 to 90 per cent in 1919, and that for steamships exceeded 100 per cent.[17] The post-First World War development of the heavy and chemical industries pushed up the proportion of metal-working and machines in total industrial output from 14 to 19 per cent in the period 1914–1919. The proportion of factory workers in those industries also increased from 11 to 17 per cent during the same period (table 1). The ratio of factory workers in the spinning and textiles industry, on the other hand, decreased from 62 to 57 per cent, though their actual number did rise. Nevertheless, reflecting the country's vigorous exports, the ratio of the textile industry's output remained at 51 per cent, the same as ten years earlier, and even showed a slight increase compared with 1914. In spite of marked advances in heavy industry during the First World War, textiles still accounted for most of the country's industry, both in output and in the number of factory workers.

The change in the industrial structure affected the composition of the female labour force (table 2). Mirroring a relative decline in the textile industry, the proportion of women workers in the entire factory workforce fell by 9 percentage points in the 1914–1919 period. Yet as of 1919 their proportion was still 54 per cent.

The content of the female workforce also changed in several ways. First of all, Article 2 of the Factory Act of 1916 (promulgated in 1911) prohibited the employment of boys and girls under the age of 12. Article 2 had a provisory clause to the effect that "the Act does not apply when workers who are over the age of ten at the time of enforcement of the Act continue to be employed." So, among existing workers it was only those under the age of ten who actually had to be taken off their jobs. The Act made it difficult to employ very young workers, and female workers under age 13, who made up not a small percentage of the workforce during the industrial revolution, quickly began to disappear (table 3).

A second change was in female labour in the textile industry, which before the industrial revolution was at the handicraft stage, centring around handlooms. Power looms were introduced in the early 1910s, and by the time the First World War broke out the industry as a whole was machine-dominated; this tendency grew even stronger with the drastic rise in textile exports during the war. The shift to power-loom factories prompted a realignment of labour. Relatively fewer workers worked in weaving, where female labour was concentrated, while the dyeing, printing, and finishing processes, in which male labour was dominant, grew in importance. As shown in "Textiles (A)" in table 2, the ratio of female workers decreased from 87 per cent in the industrial revolution period to 85 per cent in 1914, and to 82 per cent in 1919. Furthermore, because manual skills were rendered unnecessary by the power looms, the apprenticeship system collapsed. Clear evidence of this change is the transforming of the "allowance" that was at the core of the apprentice system into monthly wages. The final factor bringing about the demise of the apprenticeship system was the imposition of legal restrictions. The aforementioned Factory Act prohibiting the employment of boys and girls under 12 years old, and Article 22 of the enforcement ordinance (1916) of this law, which stipulated that "the wage for workers shall be paid in monetary form at least once every month," signalled the end of apprenticeship practices based on the employment of young children and payment in kind ("apprentice allowance" paid after the expiration of the term of service.)[18]

Another notable change during the First World War period is that the number of female workers in the metalworking and machine industry rose to over 10,000 for the first time (1916), and their percentage of that industry's workforce increased from 3–4 per cent during the industrial revolution period to more than 6 per cent in 1919. Owing to a rapid rise in labour demand caused by the brisk economy during the war, female workers were absorbed into the heavy industries sector on an unprecedented scale. A contemporary report states: "The merit of female workers is that they are more meticulous and find it less bothersome than men to do jobs that require careful work in electric motor production, such as coil winding and insulator manufacture."[19] Women workers in the machine/appliances industry were presumably employed in such jobs as required only semi-skilled work, mainly in the sector of newly emerging electric appliances. For example, in Mitsubishi's Nagasaki Shipyard, one of Japan's leading heavy industrial firms at that time, female workers increased in number from 1916 onward (449 women as of 1918), and an overwhelming proportion of these women were engaged in electric motor manufacturing.[20] Although female workers in the machines/appliances industry constituted a negligible share in numerical terms, they nevertheless deserve mention as the prototype of female labour seen later during the Second World War.

III. Inter-war Period

In the spring of 1920, Japan experienced a financial panic in reaction to the war boom earlier than in other countries, and a series of crises followed every few years. A second crisis ensued from the Great Kanto Earthquake and Fire of 1923, and another financial panic occurred in 1927 (the Shōwa Panic). The Shōwa Depression of 1930 arrived before Japan's capitalist system was able to regain relative stability. Even in the chronic recession of the 1920s, urban investment and electrification rose, facilitating the growth of the new heavy and chemical industries such as electric power, related industries (electric appliances, wire), organic synthetic chemicals, and automobiles. Despite the rapid advance of these new industries, heavy and chemical industry as a whole did not show much growth, partly because of mounting import pressures (as European products began to flow back into the Asian market after the end of the First World War), and partly because of reduced military demand stemming from the arms reduction resolutions Japan signed at the Washington Conference of 1921–1922. For example, the metalworking and machine industry produced only 19 per cent of total industrial output in 1919–1929, and the proportion of factory workers in this industry declined slightly from 17 to 16 per cent. The chemical industry showed almost the same pattern during the same period (table 1).

By contrast, the cotton and silk-thread industries were among the few thriving sectors during the chronic recession. The former was invigorated by increased cotton cloth exports to new Asian markets, and the latter by a rise in silk exports to the United States accompanying the boom in the US economy in the 1920s. As a result, though the textile industry's output ratio decreased in the late 1920s, its share of the factory labour force remained high at 56–57 per cent throughout the 1920s (table 1). The continuity of the textile industry-led industrial structure is also reflected in the high proportion of women textile workers in the total factory labour force: 53 per cent in the 1920s, about the same level as in 1919 (table 2).

Drastic change came to the textile industry-led structure following the Shōwa Panic and the Manchurian Incident of 1931 (which led eventually to the outbreak of the undeclared war between Japan and China that lasted from 1937 until the end of the Pacific War). The former struck an unprecedented blow to the silk-reeling industry, which had relied heavily on exports to the United States for growth. The decline of the industry that had been the country's greatest foreign currency earner destroyed the international balance of payments, and was partly responsible for the embargo Japan placed again on the export of gold. The embargo sharply lowered the foreign exchange rate and markedly raised the level of self-sufficiency in the heavy and chemical industries. Fiscal policy too, especially the increase in military spending, introduced by Takahashi Korekiyo (who became Finance Minister in December 1931), imparted a temporary impetus to expansion of the domestic market in the heavy and chemical industries immediately after the Manchurian Incident, leading to the subsequent growth in these industries. Moreover, expanded investment in the economic development of Manchuria brought about a sharp rise in the export to Manchuria of heavy and chemical industrial products.[21]

The above-described factors combined to cause heavy and chemical industry to emerge rapidly as a key industry in the Japanese capitalist system (table 1). In 1936 the metalworking and machine industry surpassed the textile industry in output, marking a decisive shift in Japan's industrial structure from reliance on textiles to heavy industry. Reflecting this change, the proportion of female workers in the total factory workforce fell below 50 per cent for the first time in 1933, and dropped further to 46 per cent in 1935 (table 2).

The inter-war period saw the progressive dissolution and/or transformation of the six categories of female workforce that had been created during the industrial revolution. Given the present level of research, it is impossible to give a full picture of each of the six types, so here I intend to discuss briefly the situation in four main industries: cotton-spinning, silk-reeling, textiles, and coal-mining.

1. Cotton-spinning

With the promulgation in 1923 of the revision of the Factory Act (which came into force in 1926), the delay in late-night work prohibition was shortened from a period of 15 years after the law's enforcement to a period of three years after enforcement of the revised Factory Act. This provided the incentive to rationalize the cotton-spinning industry, whose rapid growth had depended upon the late-night work of female workers. Rationalization was achieved at first by increasing the number of spindles operated by the individual worker. From around the time of the Shōwa Panic technological rationalization was sought by remodelling equipment—for example, the independent operation of electric-powered spinning machines, more rapid spindle speed achieved through better-quality rings and improved spindle designs, the shortening of the three-stage roving process through the introduction of high-draft machines, and the unification of the ginning processes into one through linked machine operations.[22] As the shift from mule to ring spindles had already taken place during the industrial revolution period, there was little room for technological improvement in the spinning machines. Technological rationalization therefore concentrated on the pre-spinning processes. In the course of the rationalization, young female workers again increased their share in the spinning industry's workforce. The proportion of women workers, which had stayed at around 78 per cent for a while until 1929, rose to 86 per cent by 1935 (table 2). The proportion of young female workers aged under 20 also increased, from 64 per cent in 1927 to 67 per cent in 1930, and then to 72 per cent in 1933.[23] This increase was notable in the jobs where there were technological improvements, such as roving and ginning,[24] prompting the replacement of male with female workers.

Also, improved technology required suitably able workers. The main concern of management gradually shifted from indiscriminate hiring to the strict screening of applicants using aptitude tests.[25] This change discouraged the traditional arrangement of hiring linked to the advance of funds that had straitjacketed workers in the past. According to a 1927 survey of 34 spinning mills, there were only five in which more than 30 per cent of the women workers had borrowed money at the time of employment. In 11 factories there were none who had done so.[26]

2. Silk-reeling

The technological rationalization that occurred in silk-reeling in the 1920s was generally partial and moderate, consisting of a nationwide standardization of silkworm species, a shift in cocoon-drying methods from heating to steaming, and division of labour between cocoon-boiling and silk-reeling. After the start of the Shōwa Depression of the 1930s, as America's demand for silk declined, there was a decisive change in silk demand from textiles to stockings, leading to the rapid spread of the multiple-spool reeler that enabled the reeling of the refined, high-quality silk required for stockings.[27] Ironically, then, in the very process of its decline, the silk-reeling industry in Japan underwent a shift from the manual-based to the machine-based factory system.

The female labour that sustained the vigorous growth of the silk-reeling industry consisted mainly of the daughters of poor tenant farmers in the 1920s, as it did during the industrial revolution.[28] Hiring practices, however, were considerably modernized, at least formally, owing to the establishment of the labour protection laws. Already during the First World War, forcing workers to pay breach-of-contract damages was banned in the 1916 enforcement ordinance of the Factory Act, which removed items regarding "advance payment" from the employment contract and made the contract simply a mutual employment agreement. After promulgation of the Factory Act revision in 1926 (Article 27 [4]), which required that the owner of a factory with 50 or more workers prepare and submit to the authorities a document laying down the employment-related regulations of the factory, labour conditions began to emerge in clearer relief.[29] The unilateral contract that had been binding only on workers disappeared, and labour contracts moved into the modern age.

3. Textiles

In the textile industry in the 1920s, as the domestic market for narrow-width cloth remained sluggish, attempts were made to shift to production of wide cloth for export, and power looms for wide cloth were introduced. From 1924 the number of narrow-cloth power looms started to decline, while that of wide-cloth looms rose throughout the 1920s. In 1925 the latter exceeded the former.[30] But only in a few regions was the shift to wide-cloth production for export using new looms successful. Many textile-producing regions did not undergo such a shift of production, and textile production grew, levelled off, or actually declined.[31] The decrease in the number of workers in the weaving industry in the 1920s (table 1) reflects this reorganization of production regions. The disintegration of the apprenticeship system that had begun during the First World War went even further in that process of change.

4. Coal-mining

The ban on late-night or underground work for women and young people included in the 1928 revision of the miners' protection regulations added impetus to a thoroughgoing technological rationalization of the coal-mining industry. Long dependent on the team labour of married couples, the new regulations forced the industry to establish new systems of production technology, and around the time of the Shōwa Depression longwall-type coal-extracting machines and conveyer equipment were introduced.[32] Coal-mining thus entered a process of change from the manual labour phase that relied on mandrels and shovels to the phase of machine extraction. The introduction of new technology put many women miners out of work. The proportion of female labour in the mine workforce had stood at 26 per cent in 1925, but decreased to 22 per cent in 1929, and to 10 per cent in 1935 (table 2).

Introduction of the new technology also destroyed the *naya* system. Joint work by many miners (made possible by the adoption of the longwall-type coal-extraction method) and mechanization of coal extraction rendered meaningless the function of the *naya* foremen as guides and supervisors of the isolated, manual work based on the married couple. The mines also required better-qualified miners than could be assembled by the indiscriminate hiring of the *naya* heads. Toward the end of the 1920s, the *naya* system was rapidly done away with in the major mines in Kyushu.[33]

Thus, with modernized employment achieved through labour protection laws and with a qualitative improvement in workers impelled by the adoption of new technologies, the types of female labour characteristic of the industrial revolution period either disappeared (as in the case of coal-mining) or underwent major changes (as in cotton-spinning, silk-reeling, and weaving) in the inter-war period, most notably around the time of the Shōwa Depression.

IV. The War Period (1937–1945)

Following the outbreak of the Sino-Japanese War in July 1937, the government tightened its control over all economic activities in the country. Wartime economic controls began with promulgation in September that year of the Extraordinary Funds Adjustment Law and the Extraordinary Export and Import Measures Law restricting investments and the purchase of industrial goods from abroad. They were made more or less complete when the National General Mobilization Law (April 1938) was enforced to mobilize both human and material resources for prosecution of the war. The main aim of the economic control was quickly to turn the industrial structure into one which, centring on munitions production, would enable Japan to overcome its relative inferiority in heavy industry productivity and continue fighting a modern all-out war. Indeed, a series of control laws, including the above-mentioned three, aided by massive government spending of public funds in the munitions industry, stimulated further progress in the heavy and chemical industries. In 1940, as shown in table 1, the proportion of metalworking and machines/appliances made up 49 per cent of total production and accounted for 45 per cent of all factory workers, whereas only 17 per cent of the production and 26 per cent of factory labour was in spinning and textiles.

The rapid advancement of the heavy and chemical industries resulted in the reduction of the proportion of women in the total factory workforce to 34 per cent in 1940. The composition of female labour, too, underwent a new, decisive change during the war period, with the rapid expansion of female labour in metalworking and machine/appliances manufacturing. The number of female workers in this sector surpassed 100,000 in 1938 and rose further to 175,000 in 1940 and over 200,000 the following year. It surpassed the number of women workers in silk-reeling in 1940 and in cotton-spinning in 1941. Also, the proportion of female labour in the total workforce in that sector rose above 10 per cent for the first time in 1939. As of March 1942,

Table 4. Number of Female Workers by Industry and Age (metalworking and machines, as of end of March 1942)

	Age 12–19	Age 20–59	Total	Proportion in total workforce
Metal-refining materials	13,868 (44.0)[a]	17,648 (56.0)	31,516 (100.0)	10.6
Other metal-working	19,394 (35.8)	34,850 (64.2)	54,244 (100.0)	15.3
Motors, machine-tools	10,505 (46.2)	12,210 (53.8)	22,715 (100.0)	11.2
Electric machines, appliances	36,432 (50.2)	36,178 (49.8)	72,610 (100.0)	24.6
Train cars, ships	7,448 (49.9)	7,486 (50.1)	14,934 (100.0)	6.5
Cars, airplanes	31,118 (52.6)	28,062 (47.4)	59,180 (100.0)	11.8
Measuring instruments, precision machinery	12,820 (52.5)	11,613 (47.5)	24,433 (100.0)	21.1
Other machine appliances	34,428 (43.4)	44,847 (56.6)	79,275 (100.0)	14.6
Total	166,013 (46.3)	192,894 (53.7)	358,907 (100.0)	14.1

a. Figures in parentheses are percentages.
Source: Health and Welfare Ministry's Labour Bureau, *Dai-5-kai rōmu dōtai chōsa kekka hōkoku* [The Report of Results of the Fifth Survey of Workers], data as of end of March 1942.

the largest numbers of women workers in the sector of metalworking and machines, as table 4 shows, were engaged in electric machines and appliances manufacturing and automobile and airplane making. Also, the proportion of women workers in each industry's workforce was highest in the electric machines and appliance industry and in measuring instruments and precision machinery, where it stood at 21–25 per cent. Presumably women workers performed relatively simple jobs that required little or no training. A notice, entitled "Re: Employment of Female Workers upon the Implementation of the Labour Mobilization Plan," sent from the Health and Welfare Ministry's Occupation Department chief and Labour Bureau chief to regional commissioners in October 1939, indicated three categories of work appropriate for women workers: "relatively easy and simple work"; "physically light work mainly using hands and fingers"; and "work that people with little or no training can perform." It listed 14 specific types of jobs suitable for women: draftsmanship, casting, lathing, turret lathing, milling, press, (small) machine assembling, finishing, electric appliance assembling, winding, insulating, electric wire wrapping, inspection, and analysis.[34] The

Table 5. Number of Female Workers by Job in 45 Machine Factories under the Jurisdiction of the Tokyo Metropolitan Police Office

Type of job	1937	1938	1939
Lathing	344	1,224	3,051 (16.4)
Inspection	1,962	3,870	3,669 (19.8)
Assembling	892	1,405	2,190 (11.8)
Recording	752	1,339	918 (4.9)
Misc. work	787	1,248	914 (4.9)
Winding	564	822	841 (4.5)
Finishing	157	311	846 (4.6)
Turret lathing	13	39	809 (4.4)
Grinding	191	356	430 (2.3)
Milling	59	168	635 (3.4)
Total (including other)	7,794	13,658	18,570 (100.0)

Source: Shōwa Kenkyūkai [Shōwa Research Society], *Rōdō shintaisei kenkyū* [Study of the New System of Labour] (Tōyō Keizai Shimpōsha, Tokyo, 1941), p. 234.

number of women working in these jobs in 45 machine factories under the jurisdiction of the Tokyo Metropolitan Police Office is shown in table 5.

The rapid increase in the numbers of women workers was a product of the labour shortage, as men were drafted in increasing numbers for the war. The series of labour mobilization policy measures designed to cope with the male labour shortage, prompting the replacement of men workers by women, is noteworthy. It is started with the above-mentioned notification of October 1939. In November 1941, a Patriotic National Labour Cooperation Ordinance was promulgated to organize patriotic labour corps (*kinrō hōkokutai*) to assist the national general mobilization campaign. Women aged between 14 and 25 were required to join the corps, and the number mobilized by this means was 1,343,000 in 1942, and 1,868,000 in 1944.[35]

In January 1943, the Cabinet agreed upon an "Outline of Emergency Measures of Labour and Production Increase," which established a quota for female employment in the types of industries and jobs where women could replace men, effectively limiting or banning male employment in these areas.[36] In September of the same year, a meeting of vice-ministers of the Health and Welfare Ministry adopted a resolution, "Re the Acceleration of the Mobilization of Women Workers," to organize female "volunteer corps" (*teishin tai*) and mobilize women for work in airplane-related factories, government workshops, and other places of work that required female labour as the result of the limiting or banning of male labour.[37]

In 1944 mobilization of women for labour was accelerated. In January, an "Outline of Emergency National Mobilization Measures" (Cabinet decision) demanded that the volunteer corps system be further activated and a target rate of female labour required by industry and job sites specified.[38] Then, with the "Outline of Emergency Measures for Decisive Battles"

Table 6. Minimum Rate of Female Employment by Industry (decided upon by the Cabinet in 1944) (percentages)

Industry	%	Industry	%
Metal industry		Airplane engine parts	45
Light metals	30	Propellers	35
Springs	50	Optical machines/tools	60
Screw	50	Firearms, bullets, other arms	40
Machine and appliances industry		Bearings	50
Ship motors	30	Valves, cocks	35
Electric machines/tools	40	Chemical industry	
Electric communications machines/tools	50	Medicine manufacture	60
Electric bulbs, tubes	60	Industrial chemicals, dyestuffs, fertilizers	20
Measuring instruments	30	Paints, pigments	30
Electric wires	40	Explosives	50
Electric batteries	50	Oil and fat	25
Machine tools	35	Rubber products	50
Tools, measuring tools	40	Leather	30
Industrial machines	30	Leather products	50
Train cars	25	Dry film plates	50
Automobiles	35	Ceramic industry	
Steel ships	20	Fire bricks	20
Airframes	40	Glass products	40
Airframe parts	40	Asbestos products	40
Airplane engines	30		

Source: Labour Ministry, *Rōdō gyōseishi* [A History of Labour Administration], vol. 1 (1961), pp. 1120–121.

adopted by the Cabinet in February and the "Outline of Measures for Strengthening the Female Volunteer Corps System" in March, women were forced to join the corps.[39] To provide the legal basis for the corps, the Women's Volunteer Work Ordinance was issued in August.[40] In the same month, the Cabinet also adopted "Controls on the Placement of Male Workers," determining the minimum female employment rate, as shown in table 6, so as to limit the hiring of men.

With implementation of the above-mentioned labour mobilization measures, a huge number of women were presumably working in heavy industries at the end of the war. Some features of heavy-industry female labour became greatly pronounced during the war. First of all, as table 4 shows, 46 per cent of female workers were under 20 years of age. In three industries, electric machines and appliances, automobile and airplane manufacturing, and measuring instruments and precision machines, more than half the female workforce consisted of girls under 20. It may also be assumed that most of the workers in the 20–59 age bracket were young women in their early twenties. According to a 1939 Health and Welfare Ministry survey, those aged 25 or more made up only 18.4 per cent of 123,000 female workers in the machines industry.[41] The high proportion of young female workers is not unrelated to the labour mobilization measures reflecting the government's basic stance in favour of the traditional family (*ie*) system. The November 1941 patriotic cooperation ordinance excluded officially or unofficially married women from those whose duty it was to join the patriotic labour corps. The January 1944 outline of emergency mobilization measures limited the mobilization of women as far as to say that "the country's family system, female characteristics, and the necessity of strengthening the national power have to be taken into consideration in promoting and expanding the labour mobilization of women." The March 1944 outline for strengthening the women's volunteer corps, too, removed "those playing a vital role in the family" from among the women to be forced to join the volunteer corps. Thus, the existence of an overwhelming proportion of young women workers must be understood in the context of the weight the government gave to the family system.

Female labour in the heavy industries during the war period and in the spinning and textile industries before the war was similar in that young workers formed the core of the workforce. They differed considerably, however, as far as academic background and social class are concerned. There are no data available concerning the educational background of women workers during the war, so we must be content to judge from the results of a 1936 survey (table 7). The survey data clearly indicate a contrast between the relatively longer education of women workers in heavy industry and the relatively shorter education of female labour in spinning and textiles. Of women workers in machines and appliances manufacturing, shipbuilding, conveyances, and precision machines, around 55 per cent were graduates of upper elementary schools (5th–7th grade) or more advanced schools. (Only 20 per cent of female workers in spinning and weaving industries had the same educational background.) Because it is unlikely that the educational background of women workers in heavy industry changed significantly after the war started (in 1937), we may assume that women workers in heavy industry in the war period were relatively better educated.

The marked difference in school education between female workers in heavy industry and spinning/textiles suggests that the former were from a higher social class. The kinds of jobs female workers in wartime heavy industry held previously or, if they had not previously held a job, the jobs of their household heads, are shown in table 8. Of those previously employed, 38 per cent had worked in manufacturing and 14 per cent in offices, totalling 52 per cent. Only 14 per cent had been engaged in farming. The two main occupations of the household heads of those who had not previously held jobs were also in manufacturing or office work, totalling 36 per cent. Farming household heads made up only 14 per cent. This indicates that a new type of female labour, which may be described as "heavy industry, urban labour," appeared as the cities supplied female labour to heavy industries during the war, as compared with the spinning and textile industries, whose main source of labour was rural villages.

Table 7. Academic Background of Female Workers, 1936 (percentages)

	Metals industry	Machines and appliances	Shipbuilding, conveyances	Precision machines	Spinning and weaving
No school education	3.9	0.9	1.5	0.4	1.2
Lower elementary school leaver	4.8	1.8	2.3	0.8	1.7
Elementary school graduate	50.2	39.3	34.3	38.0	69.3
Upper elementary school leaver	3.8	4.5	3.4	5.4	7.8
Upper elementary school graduate	31.7	42.1	31.4	43.7	17.4
Vocational schools and middle schools[a]	5.6	11.4	27.0	11.6	2.5
Total	100.0	100.0	100.0	100.0	100.0
Actual numbers	(13,851)	(23,239)	(5,306)	(6,783)	(618,900)

a. Includes those who left before completing their studies.

Source: Statistics Bureau of the Cabinet, *Rōdō tōkei jitchi chōsa hōkoku* [Report of Labour Statistics Field Survey], no. 5.

Table 8. Previous Jobs of Female Workers Employed in Metals/Machines Industries over the Past Six Months (or, in the case of workers without a previous job, their household heads' jobs)

Previous job	Previously employed	Previously unemployed (or job of household head)
Offce work	6,034 (13.6)[a]	4,405 (8.4)
Technical work	264 (0.6)	357 (0.7)
Manufacturing	16,786 (37.8)	14,500 (27.5)
Mining	158 (0.4)	195 (0.4)
Commercial	2,386 (5.4)	3,541 (6.7)
Transport, communications	632 (1.4)	479 (0.9)
Household employee	4,753 (10.7)	3,017 (5.7)
Farming	5,971 (13.5)	7,305 (13.9)
Fisheries	217 (0.5)	409 (0.8)
Other job	4,881 (11.0)	4,781 (9.1)
No job or unknown	2,283 (5.1)	13,649 (25.9)
Total	44,365 (100.0)	52,638 (100.0)

a. Figures in parentheses are percentages.
Source: Health and Welfare Ministry's Labour Bureau, *Dai-3-kai rōmu dōtai chōsa kekka hōkoku* [Report of Results of the Third Survey on Workers], data as of the end of March 1941.

V. Conclusion: Female Labour before and after 1945

The sections above outline the historical development of female labour in Japan from the industrial revolution to the end of the Second World War. The various types of female labour organized during the industrial revolution either disappeared (coal-mining) or were transformed (cotton-spinning, silk-reeling, textiles) in the inter-war period, and a new type—what I have called the "heavy industry, urban labour" type—came into being during the war period. If the range of examination is expanded to include the post-war period, this new type of female labour can be seen as a prototype of the young female labour in the metals and machine industries that emerged during the rapid economic growth period after the war. Continuity between pre-war and post-war labour patterns among women is detectable here.

However, even during the war period, when Japan suffered a severe labour shortage, middle-aged married women, in conformity with the demands of the traditional family system, did not join the labour force. The post-war reforms included the dissolution of this old family system, and the break-up of that system provided the historical premise upon which the vigorous post-war growth of Japan's economy could unfold. With the rapid growth of the consumer durable goods industry (home appliances), which produced labour-saving devices that shortened household tasks, middle-aged married women were released into the labour market in large numbers. This represents a marked discontinuity between pre-war and post-war female labour, and suggests that the social structure of a given country, the degree of its economic maturity (the degree of growth of its consumer durables industry), and the impact of these two factors on household labour have a decisive effect on the flow of women into the labour market.

Notes

1. Commerce and Industry Bureau, Ministry of Agriculture and Commerce, *Shokkō jijō* [Conditions of Workers], vol. 2 (1903), pp. 129, 186.
2. Among Japan's export items as of 1902, silk ranked first (making up 28.3 per cent of total exports), followed by silk fabrics (10.3 per cent), cotton yarn (7.3 per cent), coal (6.4 per cent), copper (3.9 per cent), tea (3.9 per cent), and matches (3.0 per cent) (*Yokohama-shi shi* [A History of Yokohama City], Shiryō-hen [Documents], part 2, 1962). Exports of straw plaiting were low that year, but the degree of this product's dependence on the foreign market (export amount divided by output) was over 100 per cent.
3. Sumiya Mikio, Kobayashi Ken'ichi, and Hyōdō Tsutomu, *Nihon shihonshugi to rōdō mondai* [Japanese Capitalism and the Labour Issue] (Tōkyō Daigaku Shuppankai, Tokyo, 1967), p. 91.
4. Takamura Naosuke, *Nihon bōsekigyōshi josetsu* [Introduction to the History of the Japanese Spinning Industry], vol. 1 (Hanawa Sholbō, Tokyo, 1971), pp. 135, 303.
5. Commerce and Industry Bureau, *Shokkō jijō*, vol. 1, p. 65.
6. Takamura Naosuke, *Nihon bōsekigyōshi josetsu*, p. 339.
7. Ishii Kanji, *Nihon sanshigyoshi bunseki* [An Analysis of the History of the Japanese Sericulture Industry] (Tōkyō Daigaku Shuppankai, Tokyo, 1972), pp. 261–264.
8. Commerce and Industry Bureau, *Shokkō jijō*, vol. 1, p. 240.
9. Ibid., p. 240.
10. Ibid., p. 285.
11. Ibid., pp. 252–260.
12. Commerce and Industry Bureau, *Shokkō jijō*, vol. 2, pp. 129, 136.
13. Ibid., p. 187.
14. Ibid., p. 188.
15. Ibid., p. 188.
16. Ibid., p. 260.
17. Takamura Naosuke, *Nihon shihonshugishi ron—sangyō shihon, teikokushugi, dokusen shihon* [A History of Japanese Capitalism: Industrial Capital, Imperialism, and Monopoly Capital] (Mineruva Shobō, Tokyo, 1980), table VIII–5.
18. For details on the collapse of the apprenticeship system in the textile industry, see Furushō Tadashi, "Ashikage orimonogyō no tenkai to nōson kōzō-'kata' no hensei to sono hōkai" [The Development of the Textile Industry in Ashikaga, and the Structure of Rural Society: Organization of "Forms" and Their Disintegration], *Tochi seido shigaku*, vol. 86 (1980): 13–15.
19. Shiraki Taiji, *Zōsenjo rōdō jōtai chōsa hōhokusho* [A Survey Report on Shipyard Labour Conditions] (a report of a school trip in summer 1991, Tokyo Higher Commercial School), p. 23.
20. Mitsubishi Nagasaki Shipyards, *Nenpō* [Annual Report] (1916–1919).
21. For more details, see Uno Kōzō, ed., *Kōza teikokushugi no kenkyū* [Studies on Imperialism], vol. 6 (Aoki Shoten, Tokyo, 1973), chap. 3, and Hashimoto Juro,

Daikyōkō-ki no Nihon shihonshugi [Japanese Capitalism during the Great Depression] (Tōkyō Daigaku Shuppankai, Tokyo, 1984), chap. 4.

22. Moriya Fumio, *Bōseki seisanhi bunseki* [An Analysis of the Cost of Cotton-spinning Production], enl. and rev. ed. (Ochanomizu Shobō, Tokyo, 1973), chap. 2, section 2; Izumi Takeo, "Taishō-ki menbō no rōdō jijō to gōrika" [Taishō-period Labour Conditions and Rationalization in Cotton-spinning], in *Senshū keizaigaku ronshū*, vol. 10, no. 2 (1976).
23. Cabinet Statistics Bureau, *Rōdō tōkei jitchi chōsa hōkoku* [Labour Statistics Field Survey Report] (1927, 1930, 1933).
24. Ibid.
25. Izumi Takeo, "Taishō-ki menbo," pp. 19–21.
26. Chūō Shokugyō Shōkai Jimukyoku [Central Employment Agency], *Bōseki rōdō fujin chōsa* [Survey of Women Workers in Spining] (1929).
27. Kajinishi Mitsuhaya, ed., *Sen'i* [Textiles], vol. 1 (Kōjunsha Shuppankyoku, 1964), pp. 471, 582–583; Kiyokawa Yukihiko, "Seishi gijutsu no fukyū denpa ni tsuite—tajōkuriitoki no baai" [The Diffusion of Silk-reeling Technology: The Case of Multiple Spool-reelers], in Hitoshibashi University, *Keizai kenkyū*, vol. 28, no. 4 (1977).
28. Yukie Miharu, "Daiichiji Taisen go no seishi jokō no sekishutsu kiban—Gokamura no nōka keiei to jokō rōdō" [The Post-First World War Supply Base of Female Labour in Silk-reeling: Farm Management and Female Factory Work in Goka Village], in Oe Shinobu, ed., *Nihon fashizumu no keisei to nōson* [Formation of Japanese Fascism and Rural Society] (Azekura Shobō, 1978), chap. 2.
29. For more details, see Oishi Kaichirō, "Koyō keiyakusho no hensen kara mita seishigyō chinrōdō no keitai henka" [Transformation of Wage Labour in the Silk-reeling Industry as Seen in the Changes in Labour Contracts], *Shakai kagaku kenkyū*, vol. 24, no. 2 (1972).
30. From *Nōshōmu tōkeihyō* [Statistical Tables on Agriculture and Commerce] and *Shōkōshō tōkeihyō* [Statistical Tables on Commerce and Industry].
31. Abe Takeshi, "Ryōtaisenkan Sennan menorimonogyō no hatten" [The Development of Cotton-weaving in Sennan during the Inter-war Period], *Tochi seido shigaku*, no. 88 (1980): 18.
32. *Nihon kōgyō hattatsushi* [A History of the Development of Japanese Mining], vol. 2 (Kōzan Konwakai, 1932), part 4, chap. 1, section 8.
33. Ichihara Ryōhei and Tanaka Mitsuo, "Tankō naya seido no hōkai (2)" [The Collapse of the Naya Mining System (2)], *Nihon Rōdō Kyōkai zasshi*, 64 (1964), pp 30–32; Tanaka Naoki, "Chikuhō sekitan kōgyō hattatsushi gaiyō" [An Outline of the History of Chikuhō Sekita Kōgyō], an essay contributed to *Asō hyakunenshi* [A Centennial History of Asō] (1975), p. 87.
34. Ministry of Labour, *Rōdō gyōseishi* [A History of Labour Administration], vol. 1 (1961), p. 927.
35. Ibid., p. 979.
36. Ibid., p. 1022.
37. Ibid., p. 1112.
38. Ibid., p. 1092.
39. Ibid., pp. 1013, 1023.
40. Ibid., p. 1089.
41. Shōwa Kenkyūkai, *Rōdō shintaisei kenkyū* [Studies of the New Labour System] (Tōyō Keizai Shimpōsha, Tokyo, 1941), p. 237.

Chapter———1

Silk-reeling Technology and Female Labour

Masanori Nakamura and Corrado Molteni

The technological development of silk-reeling during Japan's early industrialization followed a distinctive pattern. Unlike other modern industries introduced from the West, silk-reeling was characterized by a combination of indigenous Japanese technology developed during the Tokugawa period (1603–1867) and advanced Western technology. Technological development in the silk-reeling industry from the Meiji era (1868–1912) to the end of the Second World War can be divided into three major periods: (1) the early 1870s, when Western technology for mechanized silk-reeling was introduced; (2) from about 1875 to 1900, when mechanized silk-reeling was transformed and adapted to indigenous conditions and diffused throughout Japan; and (3) from the early 1900s to 1945, a period that saw major improvements and innovations in mechanized silk-reeling, such as the multiple-spool reeler. A fourth period can be designated, based on the introduction of the automatic reeling machine, but this occurred in the 1950s and is beyond the scope of this essay.

The process of introducing and adopting foreign technologies cannot be viewed separately from the historical and social conditions of the recipient country. This consideration also applies to silk-reeling in Meiji Japan. In this case, however, three factors in particular should be considered.

First, since silk-reeling was a traditional indigenous industry dating back to much earlier times, it is necessary to ascertain the technological level reached before the advent of Western technology. A second factor involves the type and features of the technology introduced and the way in which it was modified and adapted before its diffusion as a Japanese technology.

In this respect, it should be made clear that the introduction of Western technology did not mean the mere import of foreign models. When the government established its first silk filatures, it actively introduced Western technology, but took local conditions and the state of indigenous technology into account. Moreover, with the diffusion of mechanized silk-reeling from government filatures to private enterprises, the new technology was further modified and adapted to local conditions.

This process was particularly evident after the 1870s, when the development of the industry was due mainly to the efforts of middle-class farmers turned entrepreneurs. Lacking adequate financial resources, these silk-reelers of rural origin could not afford to tie up their scarce capital in costly investments for infrastructure and equipment. They needed to simplify and adapt Western technology to reduce the amount of capital required. This was the salient characteristic of technological development in Japanese silk-reeling.

Third, because of the delicacy and fineness of silk textiles, silk-reeling depended to a great extent—compared with cotton-spinning—on the dexterity and skill of the workers. Consequently, the level of technology was a determining factor in the quality of the labour force. Still, that relationship could be reversed, with the quality of the labour force determining the level of technology. This is confirmed in a 1913 survey: "In silk-reeling, machinery can be effectively applied only to a very small, limited extent. In large part, one has to rely on the skill of female workers. As a consequence, rather than rely on superior machinery, it is definitely better for efficient operations to rely mainly on the technical skill [of the workers]."[1] An earlier survey also notes that "It is as if the workers were a part of the silk-reeling machine."[2]

Of course, this peculiar feature, together with the specific type of technology adopted and its subsequent development, had far-reaching implications for the composition of the workforce, working conditions, and the organization of labour in the industry. For this reason, this essay emphasizes working conditions and labour–management relations in Japanese silk-reeling, as well as the transfer and transformation of silk-reeling technology in Japan. Before discussing these main topics, a brief explanation of the silk-reeling process itself is in order.[3]

From about 1910 to the 1920s, when the silk-reeling industry was at its height, the production of raw silk from silkworm cocoons consisted of six operations: (1) storing the cocoons; (2) cocoon-sorting; (3) cocoon-boiling; (4) silk-reeling; (5) re-reeling; and (6) silk finishing.

In the first stage, cocoons were stored in large warehouses equipped with facilities for regulating temperature and humidity. Before storage, cocoons were dried in order to kill the chrysalis and to prevent it from emerging from the cocoon as a moth. Drying also protected the cocoons from mould. Sorting eliminated defective cocoons (double-shaped or loosely formed), leaving the rest for the next operation, boiling. Here cocoons were placed in basins of hot water so that the ends of the silk filament would loosen and could easily be drawn out. The basic operation of silk-reeling consisted in first searching for and picking from the water basin the ends of the silk filaments, a process known as *sakuchō* in Japanese. Four or five of these ends were then joined and passed onto a twisting device (*yorikake sōchi*) to obtain a single thread, which was finally wound onto the reel itself.

In silk-reeling, about half of the work of the female operators consisted in monitoring the reeling thread and adding new strands whenever necessary to maintain a uniform degree of thickness. The manual addition of new filament ends to the reeling thread, an operation known as *setcho* or *tencho*, was very demanding, since the silk filaments could be easily broken, especially when the cocoon was nearly finished. Furthermore, the operation had to be executed dexterously while the machine was in motion, and the process occupied 50 per cent of the silk-reeling female labour force.[4] Needless to say, this was the most important part of the entire production process, so efforts to improve silk-reeling technology concentrated on the invention of a suitable device, a *setchoki*, that would simplify and mechanize the addition of new silk filaments.

Once reeled, raw silk was passed from the small reel frame onto a larger frame (re-reeling). The hanks thus obtained were then twirled to prevent tangling, inspected, classified, and finally packed, ready for shipment to silk markets. Below, we examine the evolution of raw silk production, focusing on the basic silk-reeling process

I. Introduction of Western Mechanized Silk-reeling

1. Traditional Silk-reeling Technology

Japanese raw silk exports jumped sharply after the opening of major ports to international trade in 1859. Before that date, indigenous silk-reeling technology had achieved a high level of development, but during the Tokugawa period it remained at the hand-reeling stage. Two slightly different reeling techniques were widespread at that time.

An 1803 document, *Yōsan hiroku* [The Secrets of Sericulture], described the *dōguri* reeling method, which involved use of a cylindrical frame. Silk filaments from several cocoons were passed through a ring of horsehair or woman's hair fixed near the water basin. The operator guided the thread with her left hand while turning with her right the cylindrical frame on which the thread was reeled. The second technique, *tebiki*, was a modification of the former. Here, the cylindrical *dōguri* was replaced by a wooden frame to which a handle was attached at one end. The operator turned the frame with her right hand. The drawback to this method was that the operator had to use the left hand to twist the filaments and could not add new filaments while reeling.

Based on the premise of household production and a self-sufficient economy, these techniques kept silk-reeling at a low level of efficiency. As a result, traditional technology was unable to satisfy the enormous growth of demand for silk following the opening of Japanese ports. After 1859, a new reeling device came into use, the *zaguri*, which could be operated sitting down. Using the *tebiki*, the reeler had to turn the frame directly with the hand. The *zaguri* was designed with a handle independent of the reel. The action of the handle was transmitted to the latter via a belt or a system of gears, accelerating the rotation of the frame. Regional variations included the Ōshū *zaguri* in north-eastern Honshu, which transmitted the rotation of the handle to the reel by means of wheels and strings. In Jōshū (present-day Gumma Prefecture), a gear wheel was used.

Zaguri reels were equipped with a swinging device that led the reeling thread from left to right and back again, allowing it to be wound evenly on the reel. This device freed one of the operator's hands from the task of changing the course of the silk thread, further improving the efficiency of the machine. The invention spread rapidly after 1859. The Jōshū *zaguri*, for example, was introduced to the neighbouring Shinshū region (present-day Nagano Prefecture) between 1860 and 1863 and quickly became ubiquitous there.[5]

But the *zaguri*, too, had its drawbacks. The operator still had to turn the reel by hand while adding new threads. The *zaguri* remained inefficient and tended to produce thread that was of uneven thickness and had many joints and knots. Its use resulted in the poor reputation of Japanese raw silk in foreign markets of the time. It was in order to overcome these shortcomings that the Meiji government introduced Western silk-reeling technology.

2. Western Silk-reeling Technology

Japan's modern silk-reeling technology came from two different sources. French technology was introduced via the Tomioka Silk Filature. A silk-reeling mill operated by a domain-run factory in Maebashi (Gumma Prefecture) and the Tsukiji Silk Filature (in Tokyo) owned by the Ono Group, a merchant house with strong government connections, imported Italian technology (table 1.1).

Of these model factories, the Tomioka Silk Filature, a fully mechanized steam-powered filature set up and operated by the Japanese government, played an important role in introducing modern technology. Shibusawa Eiichi (1840–1931), an entrepreneur and business leader during the Meiji and Taishō periods, commented on the objectives of the Tomioka plant as follows:

> At that time, the raw silk we exported was not of the same quality as the fine raw silk produced in Italy. It was all made using *zaguri* reels, none from European machines. We produced only raw silk of irregular denier [a unit for measuring the thickness of silk thread], and its utilization on the customers' markets was limited to the weft. Under these conditions, it was impossible to expand the market and make silk an important export item. We all believed we had to shift to mechanized reeling, as in France and Italy, in order to produce high-quality raw silk for the warp . . . and the first step turned out to be the establishment of the Tomioka Silk Filature.[6]

For the planning and building of the Tomioka Mill, the Meiji government employed Paul Brunat, a French technician, in June 1870. The construction

Table 1.1. Traditional and Mechanized Silk-reeling Technology (Tomioka Silk Filature)

	Tomioka S.F. (French-type)	*Zaguri seishi* (Japanese-type)
Machine	Made of metal	Wooden
Rotation of the wheel	Steam engine	Manual
Re-reeling	Required	Required
Reeling basins and cocoon-boiling basins	Separated	Combined
Cocoon-boiling	Steam	Direct heat
Operator	Performed both operations of cocoon-boiling and silk-reeling proper	Performed all operations including reel rotation

Source: Okumura Shōji, *Koban, kiito, watetsu* [Gold Coins, Raw Silk, Japanese Iron] (Iwanami Shoten, Tokyo, 1973), p. 107.

of this modern silk filature was completed about two years later at a total cost of 198,572 yen. The Tomioka plant was a large-scale, Western-style complex built of red brick. It comprised a total of 17 buildings, including cocoon warehouses, a boiler room, building for cocoon-drying, silk-reeling, and re-reeling, dormitories for the workers, and an official residence for the French employees. The facility was equipped with 300 reeling basins, making it a significantly larger factory than the Maebashi Silk Filature (12 reeling basins) or the Tsukiji Silk Filature (60 reeling basins). According to a newspaper report of that time: "If we should mention the three largest enterprises in our country today, these would be the Osaka Mint, the Yokosuka Shipyard, and the Tomioka Silk Filature. These are enterprises rarely found in any European country and even Westerners often praise them."[7]

In what way, then, did the technology of the Tomioka Silk Filature differ from traditional Japanese *zaguri* technology? As shown in table 1.1, the basic difference between traditional and Western (Tsukiji and Tomioka Silk Filature) silk-reeling technology concerned:

1. The method of rotating the reeling frame. This was completely mechanized through the substitution of manual operations by a steam-powered mechanism. It was thus possible for the operator to use both hands for the silk-reeling process itself, a radical change from the traditional reeling method.
2. The method of heating water for cocoon-boiling. Here steam produced by a central boiler was used for heating the water in all the basins instead of the traditional method of heating by means of small fire placed directly under each basin.
3. Separation of the water basin for cocoon-boiling from the basin for silk-reeling itself.

4. The application of a device for twisting the silk thread. In this way, various silk filaments were intertwined, while, through friction, water was squeezed out and the remaining sericin acted as a binding agent combining all filaments in a single thread.[8]

There were other innovations involving operations other than the silk-reeling itself, for example, in the method employed for killing the chrysalis and drying the cocoons.

> During the Edo period, fresh cocoons were exposed to the sun, whose warmth killed the chrysalis. This was the so-called sun-drying method, but Brunat did not use it and adopted instead the method of killing the chrysalis by steam. The sun-drying method lowered the quality of the cocoons. Moreover, if there was insufficient sunlight the chrysalis was sometimes not killed and the moth or maggot could develop in the cocoon. [However], with the new method using steam the cocoons placed in a container were exposed for a short time to a hot steam current and were then immediately dried. After the establishment of the Tomioka Silk Filature, silk-reelers everywhere in Japan changed at once to the new method.[9]

As shown above, the silk-reeling technology of the model factories of Tsukiji and Tomioka marked the passage from the tool stage to the machine stage, although, as already pointed out, silk-reeling machines were still imperfect devices, since the crucial operation of adding new threads (*setcho*) had still to be performed manually by the operator. For this reason, it is more correct to say that the Western type of silk-reeling machine represented an intermediate stage of development between production tool and machine. It is also important to note that French silk-reeling technology was not introduced passively, but the equipment and the production process adjusted and modified to suit the specific conditions in Japan, the quality of its cocoons, and the level of existing traditional technology.[10] For example, in Europe, silk thread was reeled directly on a large frame, but in Tomioka the process was separated into two stages. First, silk was reeled on a small frame and only later re-reeled onto a larger one.

While the large factories introduced up-to-date machines and other equipment relatively smoothly, the recruitment of female workers posed the greatest difficulty for the management of the Tomioka Silk Filature. Since only five or six years had elapsed from the fall of the Tokugawa regime, prejudices and feudal customs still survived among the people. A feeling of distrust of and opposition to the adoption of modern (Western) civilization and "enlightened ideas" was widespread. The Meiji government first announced the recruitment of female workers in the Iruma area near Tokyo, as well as in Gumma, Tochigi, Saitama, and Nagano, but the number of applicants turned out to be a mere 50 or 60. Resistance and rumour spread: "It is unthinkable that we should be taught by Westerners!" "The smoke of the coal is poisonous." "They [the Westerners] extract blood and drink it every day [referring to red wine]." It was thus not surprising that the number of persons willing to apply was very small. Thereafter, in May 1872, the government distributed throughout the country a pamphlet on silk-reeling, *Seishi kokuyusho* [Official Proclamation on the Silk-reeling Industry] and in September of the same year issued in ten northern prefectures (Mizusawa, Iwate, Miyagi, Akita, Iwaki, Yamagata, Wakamatsu, Fukushima, Oitama, and Sakata) "Regulations concerning the Employment of Female Trainees in Silk-reeling at the Tomioka Silk Filature" (*Tomioka seishi kōjō kuriito denshūkōjō yatoiire kokoroe*). Notwithstanding these efforts to quell the rumours, the number of applicants did not increase. In these circumstances, the government had no choice but to resort to the employment of daughters of the old class of samurai, but even so it could not recruit more than 100 workers, a fourth of the workforce actually needed. At this point, the first factory director, Odaka Atsutada, convinced all the officials involved with the project to send their daughters to the factory. This pump-priming policy proved effective, and the number of applicants began to increase, finally reaching 404 in February 1873. This process is vividly described in *Tomioka nikki* [Tomioka Diary] by Wada Hide, the daughter of a samurai from Nagano Prefecture who was among the first employees at the Tomioka Silk Filature. According to the register of the dormitory of Tomioka Silk Filature, female workers were recruited from all over the country. The majority, 278, were from Gumma Prefecture, where the factory was located, but others came from Iruma (98), Nagano (11), Tochigi (5), Tokyo (1), Nara (2), Mizawa (8), Oitama (14), Miyagi (15), Shizuoka (6), Hamamatsu (12), Sakata (3), and Ishikawa (1). Almost all were young, unmarried girls of about 18 years old, who later made an important contribution to the diffusion of mechanized silk-reeling technology in other parts of the country.

The case of the Tomioka Silk Filature is interesting, as it shows how a new technology introduced into a different social environment can produce all kinds of cultural frictions. On the other hand, the success of the enterprise produced the stimulus for the diffusion of technical and organizational knowledge concerning mechanized silk-reeling. It also contributed to the formation of the first group of skilled workers.

3. Transformation and Adaptation of Western Silk-reeling Technology

The influence of the Tomioka and Tsukiji silk filatures soon spread throughout the country. From the end of the 1870s, filatures operating with steam were set up in rapid succession, especially in Nagano, Gifu, and Yamanashi prefectures in central Japan. The industry flourished particularly in Nagano (then known as Shinano), where in 1879 there were already 358 steam filatures employing more than ten workers, that is, 54 per cent of the total of 655 filatures nationwide (see table 1.2).

According to the second volume of *Shinano sanshigyo-shi* [History of

Table 1.2. Diffusion of Western Silk-reeling Technology in the Early Meiji Period (factories with 30 or more reeling basins)

Year established	Prefecture	Founder	No. of basins	Power source
A. French type (Tomioka Silk Filature model)				
1874	Nagano	Ōsato C. (Rokkusha)	50	Hydraulic
1874	Nagano	Takahashi H. (Seishisha)	32	Hydraulic
1874	Ishikawa	Prefectural Silk Filature of Kanazawa	100	Hydraulic
1874	Tōyama	Hirano I. et al.	64	Hydraulic
1877	Hyōgo	Prefectural Model Silk Filature (Kakusansha)	76	Hydraulic
B. Italian type (Maebashi Silk Filature model)				
1872	Gumma	Hoshino C. (Mizunuma Silk Filature)	32	—
1875	Gumma	Fukazawa O. and Kuwashima S. (Sekine Silk Filature)	48	—
1876	Saitama	Shimizu M. (Yōgyōsha)	80	Hydraulic
C. Italian type (Tsukiji Silk Filature model)				
1872	Nagano	Dobashi H. (Miyamada Silk Filature)	36	Hydraulic
1873	Nagano	Hirazawa C. et al.	54	Hydraulic
1873	Nagano	Kamiya N.	72	—
1873	Nagano	Yamagishi S. (Shin-yōsha)	200	—
1873	Fukushima	Anzai U. (Nihonmatsu Silk Filature)	48	Hydraulic
1873	Fukushima	Kawamata Silk Filature	48	Hydraulic
1875	Nagano	Shiozawa S. (Taiyōsha)	32	Hydraulic
D. Mixed type: French and Italian technology				
1873	Nagano	Seki K.	96	Hydraulic
1873	Yamanashi	Prefectural Silk Filature	30	—
1873	Niigata	(Gosen Silk Filature)	60	—
1874	Nagano	Hasegawa H.	30	Hydraulic
1875	Nagano	Susaka Y.	80	Hydraulic
1876	Nagano	Okuzawa K.	60	—
1876	Nagano	Kitahara S.	40	Hydraulic
1876	Yamanashi	Kazama I. (Seishinsha)	60	—
1877	Nagano	Katakura K.	32	Hydraulic

Source: Katō Sōichi, *Nihon seishi gijutsu-shi* [History of Silk-reeling Technology in Japan] (Seishi Gijutsushi Kenkyūkai, 1976), pp. 105–109.

Sericulture in Shinano], Italian silk-reeling technology from the Tsukiji Silk Filature was introduced at first in the Miyamada Silk Filature located in Suwa district, in the southern part of Nagano Prefecture. On the other hand, French technology from the Tomioka Silk Filature was introduced at Rokkusha, a company set up in Matsushiro, in the northern part of the prefecture. Further, at Nakayamasha, a silk filature established in 1875 in the town of Hirano (present-day Okaya) in Suwa, French and Italian technology were combined in the so-called "Suwa method" or "Shinshū method."

As for Rokkusha, it started operations on 22 July 1874, with 50 reeling basins. Among its employees were Wada Hide and 15 other female workers who had learned the new skill required for mechanized silk-reeling at the Tomioka Silk Filature. To them the company entrusted the task of instructing and training the other workers. Although the company had taken the Tomioka Silk Filature as its model, it suffered as a private enterprise from a shortage of capital, and for this reason its founders attempted to simplify and reduce the cost of the factory's facilities and equipment. When Wada Hide visited the factory in July 1874 upon returning from Tomioka, she recorded in her diary the following impressions:

> I saw the machinery and the rest. As I was already prepared, I did not feel particularly surprised. On the contrary, I thought that they had managed to do it fairly well. However, as regards the difference with Tomioka, it was like comparing heaven and earth. Copper, iron, and brass had become wood. Glass had changed into wire, bricks into clay. It was like seeing a dream within a dream. Well, that day I left thinking that, after all, their having managed to start using steam to get the silk was praiseworthy, at least.[11]

This simplification of Western technology described by Wada was carried even further in the case of Nakayamasha, a company set up by nine small silk-reelers led by Takei Daijirō and equipped with 100 reeling basins. As already mentioned, the Nakayamasha blended the French and Italian technology of the Fukayamada and the Rokkusha, thus developing the prototype of the "Suwa method," for mechanized silk-reeling. Since this method was widespread in Japan, six features that distinguished Nakayamasha from the Tomioka Silk Filature are listed below.[12]

1. As in the case of Rokkusha, wooden structures were substituted for all the Western-style brick edifices.
2. The power source for rotating the reels had been changed from steam to water, although later, in other silk filatures, even human power was employed. Elsewhere, the history of power sources in silk-reeling followed the general pattern of development: from human to water, steam, and finally electric power. Actually, throughout the Meiji period human power, in the direct form of manual operation of reels and more indirectly by means of treadles, continued to be relied upon in silk-reeling, together with water and steam power. On the other hand, in the Nakayamasha and other steam filatures adopting the "Suwa method,"

steam boilers continued to be used to produce the energy for heating the water in reeling basins.

3. The material for the reeling machines was changed from metal to wood.
4. The material and the shape of the reeling basins was changed, the former from metal to ceramic, and the latter from a circular to a semi-circular shape. The new semi-circular reeling basins were not only less costly to produce, they were even superior to the cooper-made ones since they did not damage the brilliance of the silk.[13]
5. Twisting devices made of copper and zinc were changed to wood and a new "Inazuma" type of device created by combining the French and Italian models. With the Inazuma the thread was frail and not well twisted, but the device was easy to handle and efficient in terms of productivity, so its adoption quickly spread through the Suwa district and throughout Nagano Prefecture. The Inazuma twisting device also featured a mechanism to prevent friction caused by back-up of the thread, helping to decrease the number of thread breaks and thereby enhancing labour productivity. Its drawback was that the threads did not hold together well, lowering the quality of the thread. The new "Azuma-type" twisting device invented in 1896 succeeded in eliminating these faults in the Inazuma, and its fundamental principle continued to be used even after the application of the multiple-spool reeling machine.
6. Modifications had also been introduced in the method for picking out the ends of the silk filament (*sakucho*) and for adding new ends to the reeling thread (*setcho*). In the former operation, a small brush made of *migo* (the core of the ears of the rice plant) came to be widely used. In the case of *setcho*, the practice adopted was to "throw" new ends to the reeling thread in order to join them. This practice, known as *nagetsuke* in Japanese, was particularly suitable for large-scale, low-quality production. It became widespread in Nagano Prefecture, where it continued to be used until the end of the nineteenth century.

As we have seen, the technical changes introduced at the Nakayamasha, and then diffused as the "Suwa method" of silk-reeling, represent a typical example of the adaptation to Japanese conditions of a technology originally introduced from the West. What is important to note is that this process meant simplification of Western technology, although it implied some improvements as well. It was determined by the need to save scarce capital resources and to reduce production costs, and it did result in higher relative output, although this was accompanied by a definite lowering of standards in the quality of the raw silk produced.

The new indigenous type of silk-reeling technology became widespread. With the exception of a minority of silk-reelers producing high-quality silk for the warp, it formed the backbone of Japanese silk-reeling technology as late as the end of the 1890s. Meanwhile, the production of machine-reeled raw silk continued to grow and in 1894 it surpassed in quantitative terms the production of raw silk reeled by the traditional *zaguri* method. This clearly shows that mechanized silk-reeling was by that time well established in

Japan. It was on the basis of this acquired strength that, from the turn of the century, Japanese silk-reelers began to improve the technology they had inherited from the second half of the 1870s.

However, before analysing this new phase of technical change in silk-reeling, it might be useful to examine more closely how Japanese silk-reelers faced the problems accompanying technological progress.

4. Silk-reelers' Attitudes toward Technological Progress

By the turn of the century, the Japanese silk-reeling industry had acquired a productive capacity competitive with that of Italy, Japan's major competitor in the American market. However, the type and the structure of reeling machines had not basically changed, nor could any substantial technical progress be noticed. For example, in 1905, silk-reeling machines completely built of iron, common in advanced countries, represented a mere 4.1 per cent of the machinery in Japan. Of these machines, 68.5 per cent were still made entirely of wood.[14] Of course, this was linked to the backwardness of both the iron-and-steel and the machine industries in Japan, but it was also related to the passivity of silk-reelers toward technical progress.

As late as 1931, for example, Harada Shinichi, the inventor of the automatic reeling machine, lamented: "I would like to try the automatic silk-reeling machine I invented, but the silk-reeling industry is more than 50 years behind. Silk-reelers, too, have in mind only immediate interest rates and prices. Moreover, circumstances are such that capital is badly needed and there is no way to introduce more advanced machinery."[15] This attitude toward technological progress was a constant feature of the Meiji period, as shown by the following examples.

In the early years of Meiji, the Japanese government had sent technical trainees abroad to have them learn French and Italian sericulture and silk-reeling technology. Tanaka Bunsuke and Sasaki Nagaatsu were two leading individuals within this group of technical trainees. In his later years, Tanaka recalled that:

> In 1886 I went to France, where I obtained a set of devices for adding new silk threads in four-spool reeling machines. After bringing it back to Japan, I experimented with the new device, but . . . without satisfactory results. After all sorts of considerations, I invented and applied to the original device a thread-cutter and a comber for eliminating knots and joints. In 1889 I obtained a patent and then tried to spread its use in the industry. [However], at that time silk-reelers were mostly satisfied with two-spool reeling machines and they did not like three- or four-spool reeling machines. Thus they were extremely cool toward the new device. . . .[16]

Why, then, were silk-reelers so unenthusiastic about the new device? On this point Onda Sadao and Tōno Denjirō observe in *Seishi shinron* [New

Study of Silk-reeling]: "At present the tendency among Japanese silk-reelers is to view the use of the *setchoki* (device for adding new filament ends while the machine is in motion) as they would a destitute person wearing rags who displays a gold watch hung round his neck. It is not an exaggeration to say that today, even if we should use the *setchoki*, its benefits would be practically nil."[17]

The reason lay in the fact that the *setchoki* was either a delicately built device easily subject to breakdown, or, if sturdily built, a device whose use required even more time than manual operation. Besides, the application of the costly *setchoki* was also considered a loss, since Japan could take advantage of cheap female labour highly skilled in manual operation. The authors of *Seishi shinron* had a thorough knowledge of actual conditions in the private silk filatures. They were well aware that there was a gap, which could not be easily filled, between the technology directly transferred from abroad and the technology of private enterprises. No matter how extraordinary the new inventions, their application and diffusion was very difficult, unless the conditions for reception were already present in the private sector. As mentioned before, these conditions developed in Japan by around the turn of the century, at first in the larger enterprises but gradually throughout the industry.

5. Improvement and Innovation in Mechanized Silk-reeling

With the turn of the century, the silk-reeling industry in Japan, which had maintained a conservative attitude toward technological progress, began to introduce improvements and innovations in technology. According to the research of Ishii Kanji, until the end of the 1890s the majority of Japanese silk filatures were equipped with simplified versions of the two-spool reeling machine. In contrast, in Italy and France technological progress in silk reeling had advanced conspicuously. Four- to six-spool reeling machines were already widely used, while in the 1910s six to eight spools were standard.[18]

Labour productivity in Japanese silk-reeling was also low. It is estimated that the annual production per capita of female Japanese workers was roughly half the production of Italian and French workers.

However, after the beginning of the twentieth century, three-spool reeling machines became widespread in the Suwa district in Nagano Prefecture. Then, at the end of the first decade of the twentieth century, the Katakura Group (Japan's largest silk-reeling company) and other major silk-reelers took the lead in the introduction of four-spool reeling machines, which rapidly spread in use. By 1911 four-spool reeling machines represented more than half the machinery installed in Japan.

From the turn of the century, the number of small silk filatures equipped with fewer than 50 reeling basins began to decline as well. Conversely, larger enterprises with 100 or more reeling basins became more numerous. In particular, during this period some very large enterprises emerged, equipped with 500 or more reeling basins.

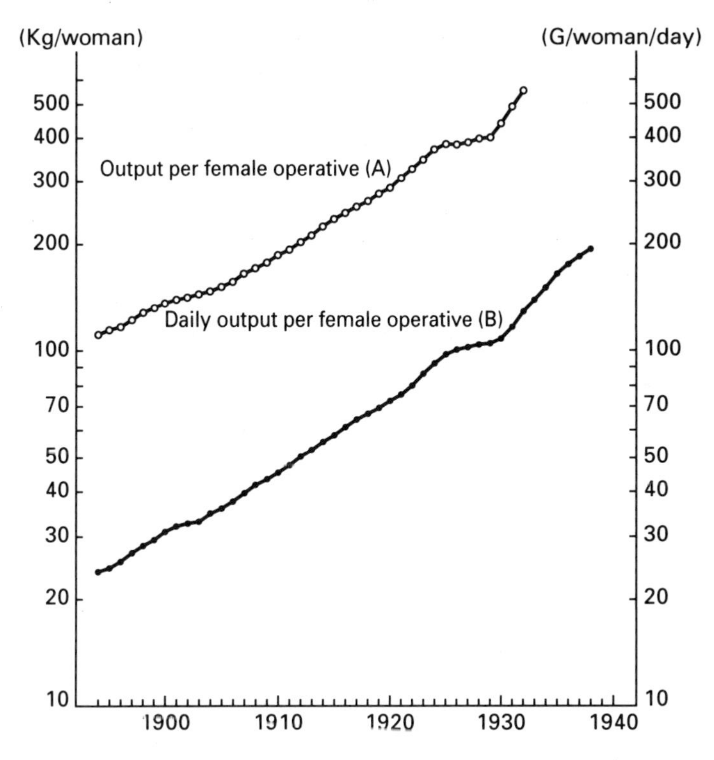

Fig. 1.1. Labour Productivity in Silk-reeling (five-year moving average)
Note: Silk filatures with ten or more reeling basins. Figures are on a silk-year basis: June to May.
Source: Fujino Shōzaburō et al., *Sen'i kōgyō* [The Textile Industry], vol. 11, in Ōkawa Kazushi et al., ed., *Chōki keizai tōkei* [Estimates of Long-term Economic Statistics of Japan since 1868] (Tōyō Keizai Shimpōsha, Tokyo, 1979), p. 169.

As a result of these developments, annual and daily worker per capita production of raw silk increased steadily, as shown in figure 1.1. Moreover, the quality of Japanese raw silk was also improved, so that by 1910 Japan had become the major exporter of raw silk on world markets, surpassing Italy, China, and France. (Seventy per cent of Japanese exports went to the American market.)

Technical progress also continued during the Taishō period, when it affected sericulture as well as reeling. Later on, a major innovation was made with the application of the multiple-spool reeling machine. The introduction of this machine permitted a conspicuous increase in labour productivity and an improvement in the quality of the silk, since it enabled further mechanization of the production process and significantly reduced the artisan character of silk-reeling.

The multiple-spool machine had been conceived and tested by Minorikawa Naosaburō, a Japanese inventor, as early as 1903, but it did not find practical application until much later. In 1919 a new type of *setchoki* (the V-shaped revolving *setchoki*) was invented, which made the multiple-spools reeling machine even more efficient.[20] The device was first introduced by the Katakura Group and other large silk filatures only in the 1920s. It came into general use in the subsequent decade, when Japanese silk-reeling, suffering from the depression of the American market and the diffusion of artificial silk, had already begun to decline. Thus, from the second half of the 1870s until the 1920s, Japanese silk-reeling production continued to be basically dependent on the skill and performance of female workers. As we shall see in the next section, in order to be competitive on international markets, Japanese silk-reeling concerns continued to rely for their capital accumulation on low wages, harsh working conditions, and a pre-modern type of industrial relations.

II. Female Labour in Silk-reeling

1. Composition of the Workforce

This section examines the composition of the workforce in Japanese silk-reeling, particularly in the silk filatures around Suwa (Nagano Prefecture), where the industry was concentrated during the Meiji and Taishō periods. In general, the number of workers in silk-reeling grew considerably from the middle of the 1880s, the years of the onset of the Japanese industrial revolution. As shown in table 1.3, in 1879 female workers in Japanese silk filatures employing ten or more workers numbered fewer than 18,000, but their numbers rose to almost 125,000 by 1896. Then, after a brief stagnation toward the end of the century, the number of female workers grew constantly, reaching a peak of more than 375,000 in 1929.

Following the distribution patterns of the industry, workers were particularly numerous in Nagano (about one-third of the total) and specifically in the Suwa district, around the lake of the same name (table 1.4). A large number of silk-reeling workers could also be found in Gifu and Yamanashi prefectures in central Honshu. Toward the end of the Meiji period, there was a conspicuous increase also in Aichi Prefecture (Tokai area) and in the northern part of the Kanto Plain (Gumma and Saitama prefectures), partly as a result of the establishment of new silk filatures by reelers from Suwa.

As already pointed out, female workers comprised the great majority of workers in the sector. For example, in 1911 there were only 1,884 male workers employed in the silk filatures of Suwa as compared to 23,445 females.[21] In general, male workers in silk filatures were assigned only to supervisory tasks, or to such jobs as cocoon drying and storage, packing, and transport of finished products. The main operations of cocoon-reeling and re-reeling were invariably performed by female workers.

Table 1.3. Female Workers in Silk Filatures with Ten or More Employees, 1879–1934

Year	Silk filatures (A)	Number of female workers (B)	B/A
1879	666	17,084	25.7
1896	2,283	124,441	54.5
1900	2,072	117,861	56.9
1905	2,320	126,077	54.3
1908	2,385	159,460	66.9
1911	2,491	191,855	77.0
1915	2,260	206,650	91.4
1918	2,680	284,549	106.2
1921	2,693	293,815	109.1
1926	2,789	312,792	112.2
1929	3,352	375,330	112.0
1934	2,709	261,344	96.5

Source: For 1879, Ishii Kanji, *Nihon sanshigyō-shi bunseki* [Analysis of the History of the Japanese Silk Industry] (Tokyo Daigaku Shuppankai, Tokyo, 1972), p. 129; 1896–1921, calculated from Nōshōmushō Nōmukyoku [Ministry of Agriculture and Commerce, Agricultural Affairs Bureau], *Zenkoku seishi kōjō chōsahyō* [Survey of Silk-reeling Factories throughout Japan], nos. 3 and 9 (1902 and 1923); 1926–1934, Nōrinshō Sanshi Kyoku [Ministry of Agriculture and Forestry, Sericulture Bureau], *Sanshigyō yōran* [A Survey of the Silk Forestry] (1935), p. 88.

Table 1.4. Distribution of Female Workers in Silk-reeling

Area	1900	1911	1912
Nagano Pref.	32,813	56,289	85,186
Suwa district	11,180	23,445	36,667
Gifu Pref.	9,886	13,187	12,321
Yamanashi Pref.	7,271	14,099	12,054
Aichi Pref.	6,062	18,286	19,903
Shizuoka	5,835	6,260	5,921
Yamagata	3,400	5,831	8,289
Gumma Pref.	3,019	6,588	25,110
Saitama Pref.	1,386	10,344	18,111
Fukushima Pref.	1,637	4,141	10,492
Other	44,102	56,830	96,428
Total	117,861	191,855	293,815

Source: Calculated from Nōshōmushō Nōmukyoku [Ministry of Agriculture and Commerce, Agricultural Affairs Bureau], *Zenkoku seishi kōjō chōsahyō* [Survey of Silk-reeling Factories throughout Japan], nos. 3, 6, and 9 (1902, 1913, and 1923).

Table 1.5. Age of Female Workers in Silk-reeling, Suwa District, May 1923

Age (years)	Number of workers	Percentage
<13	150	0.4
13–14	5,811	14.9
15–17	12,682	32.5
18–20	9,595	24.6
21–25	7,812	20.0
26–30	1,760	4.5
31–40	955	2.4
41–50	266	0.7
More than 50	11	—
Total	39,042	100.0

Source: Calculated from *Nagano kenshi, Kindai shiryōhen, sanshigyō* [History of Nagano Prefecture, Documents on the Modern Period, Silk Industry], vol. 5. (3) (1980), p. 898.

The average age of female workers was extremely low. According to a survey of 205 silk filatures in Nagano Prefecture in 1901, 66 per cent of the workers were under 20 years old. Of these, 2,285 (18 per cent) were under 14, and 150 were little children of less than 10 years old.[22]

In 1916, the enforcement of the Factory Act (Kōjōhō) restricted to a certain extent the exploitation of child labour in silk-reeling, but its effects were rather limited. In principle, the Factory Law prohibited the employment of workers less than 12 years old, but this clause applied only to factories with 15 or more workers or to factories whose environment could be detrimental to the health of child workers. Besides, administrative officials could always authorize the employment of children of 10 or older, "provided that they were assigned to simple, light tasks."[23]

As shown in table 1.5, as late as 1923 the majority of the workforce in Suwa's silk filatures still consisted of very young female workers of less than 20 years old. In general, most of the girls left the factory when reaching the age then considered suitable for marriage.

As for the origin of the Suwa workers, in the early stages of the industrialization process female workers were mostly recruited locally, in the villages of the Suwa district or in the neighbouring districts of Ina and Higashi-Chikuma.

In the first place, the low percentage of female workers coming from the Suwa district in subsequent years should be noted. This percentage in 1903 was already a mere 12.6 per cent, and it decreased further to 8.4 per cent in 1918. The percentage of workers from the neighbouring districts of Kamiina, Higashi-Chikuma, and Gifu Prefecture decreases as well,[25] while, on the contrary, the number of workers from faraway districts or prefectures

Table 1.6. Origin of Female Workers in Okaya Silk Filatures[a]

Area	1903		1918	
	Number of workers	%	Number of workers	%
Nagano Prefecture				
Suwa district	521	12.6	2,063	8.4
Kami-Ina	792	19.2	2,113	8.6
Shimo-Ina	98	2.4	184	0.8
Nishi-Chikuma	4	0.1	473	1.9
Higashi-Chikuma	690	16.7	2,795	11.4
Minami-Azumino	161	3.9	1,027	4.2
Kita-Azumino	145	3.5	703	2.8
Minami-Saku	5	0.1	341	1.4
Kita-Saku	49	1.2	423	1.7
Chiisagata	117	2.8	824	3.4
Other	36	0.9	504	2.1
Hanishina	44	1.1	920	3.8
Sarashina	27	1.1	759	3.8
Kamiminochi	—	—	786	3.1
Kamitakai	—	—	674	2.7
Total	2,689	65.2	14,589	59.5
Other prefectures				
Yamanashi	670	16.3	4,541	18.5
Niigata	—	—	2,434	9.9
Toyama	195	4.7	1,391	5.7
Gifu	592	13.6	1,233	5.0
Other	9	0.2	338	1.4
Total	1,436	34.8	9,937	40.5
Grand total	4,125	100.0	24,526	100.0

a. The town of Okaya (then known as Hirano) was the most important centre of silk-reeling in the Suwa district.
Source: Calculated from Hirano-mura Yakuba [Hirano Village Hall], *Hirano sonshi* [History of the Village of Hirano], vol. 2 (1932), p. 406

shows a remarkable increase. In particular, an increasing number of workers was recruited from the prefectures of Yamanashi and Niigata, accounting in 1918 for 28.4 per cent to the total of female employees in the Okaya filatures.

The regions supplying female workers to the silk-reeling industry were agricultural areas with almost no industry (table 1.6). As pointed out by Ishii Kanji, these regions were characterized by the existence of a high percentage of tenant farmers—higher than in the rest of Japan—or, as in the case of the northern part of Gifu Prefecture, by an extremely low level of agricultural productivity.[26] Moreover, as shown by the research of Ishii Kanji and Nakamura Masanori, the vast majority of female workers came from farmers' families belonging either to the group of tenant farmers or to the group of small proprietors owning a tiny patch of arable land.[27]

For tenant farmers, the small income brought home by their daughters working in silk filatures in distant regions was badly needed to make a living possible and to pay the exorbitant, oppressive tenant fees, which could amount to 50 or 60 per cent of the yearly crop.[28] On the other hand, it was probably because they were so inured to oppression at the hands of local landlords that these workers from rural areas found even the extremely harsh working conditions of the silk filatures bearable. In this sense we can say that the landlord system which dominated Japan's pre-war rural society was closely linked to the development of silk-reeling.

From the above analysis we may conclude that, during the second half of the Meiji period and in the Taishō period, the composition of the workforce in silk-reeling demonstrated the following characteristics:

1. The overwhelming majority were young and female workers.
2. The period of employment was limited to a few years before marriage.
3. Workers came from rural districts—in particular from families of tenant farmers or small owner-farmers—often located in remote areas far from the centre of industry.

To sum up, silk-reeling workers continued to be closely linked with the traditional rural society in which they were born and to which they usually returned after leaving the factory. In this respect, they clearly belonged to the *dekasegi* type, that is, workers closely tied to their native villages but whose temporary occupation was in the industrial sector.[29] This peculiarity of the workforce in silk-reeling was further reinforced by the seasonal character of the industry: operations were usually suspended during the winter, forcing all workers to return to their villages. As a whole, then, it was a docile workforce which could be easily controlled and regimented, although episodes of passive and even active resistance against harsh working conditions did occur, especially during the 1920s.

2. Labour–Employer Relations

We shall turn now to the analysis of the employment practices in silk-reeling, and in particular to the recruitment system and the employment contract. It should be noted first that the contract of employment lasted for only one season, and the hiring of workers for the silk filatures of Suwa was repeated every year during the winter months, when operations were suspended. The hiring was done by recruiters (usually the male supervisors in the silk filatures), who were sent to the remote rural areas supplying the workers. Recruiters relied on the service of local middlemen who provided introductions to families of potential workers. This is the way the vast majority of workers were recruited, since at that time a horizontal, free labour market had not yet developed. For workers in silk-reeling it was actually rather difficult to move from one factory to another owing to the existence of a rigid, binding employment relationship. Worker mobility was made even more difficult by the practice of accommodating workers in factory dormitories and, from the beginning of this century, by the introduction of a system for the registration of workers.[30] Nevertheless, as we shall see later, worker mobility could not be completely controlled, especially in times of sharp increases in labour demand. In these periods, silk-reelers could even resort to illegal, violent means in order to ensure a sufficient number of workers. According to *Shokkō jijō* (a government report on the conditions of factory workers published in 1903), "at the time of growing demand for raw silk . . . [silk-reelers] in Suwa used thugs to kidnap girls employed by other factories during their outings. . . ."[31]

As for employment contracts, we quote below an example of a typical contract used by the silk-reelers of Suwa. This contract was signed in 1895 between the head of the family of a female worker from the district of Higashi-Chikuma (Nagano Prefecture) and the silk filature of Miyazaka Kauemon, in Hirano (present-day Okaya:)

Agreement concerning Female Workers
in Silk-reeling

Amount: 3 yen

Worker
(Address)
(Name)

I declare hereby that I have duly received the sum of 3 yen on the understanding that the above-mentioned person will be employed in silk-reeling as a worker in your "house" in the —— year of Meiji until December of the same year. It is understood that she will work without fail during the period, and she will strictly respect the established customs of your "house." In case, by unavoidable circumstances, she might not be able to work, in order to avoid any inconvenience, a substitute shall be immediately found and sent to you. It is understood that during the time of the agreement she will not work at other places.

In case of transgression of the above agreement a sum equal to —— times the amount received will be paid as damages.

28 January 1895
(Name of family head)
(Name of worker)

(Address and name of
the company)

Several features of this "agreement" are notable. In the first place, it should be noted that the contract did not take the form of a bilateral contract of employment, but was drafted in the form of a unilateral promise on the part of the worker. No mention is made of conditions of employment such as wages, working hours, and so on. The unilateral promise was based simply on the receipt of a sum of money advanced by the employer, who thus obtained the right to employ the worker for the period of about one year.

A second point of importance is that the duties of the worker were only vaguely defined; they were not limited to the execution of reeling work, but even included respect for the established "customs" of the house. Third, the worker was not allowed to leave the factory during the time of the contract. A substitute had to be found in case unavoidable circumstances prevented her from working. Fourth, in case of transgression of one of the clauses of the contract, compensation had to be paid—compensation which could amount to 20 times the amount of the money advanced by the employer. Finally, it should be noted that the contract was stipulated and signed by the head of the family. This was partly due to the young age of the workers, but more generally it reflected the fact that in Meiji Japan the family had absolute authority over the members of the household and particularly over female members, who were treated virtually as incompetents.

In sum, female workers in silk-reeling were not hired as free wage-earners; their condition closely resembled the state of bonded feudal serfs, whose obligation in this case originated in the sum of money advanced by the employer.

Later, with the turn of the century, contracts of employment became more formal, in line with the requirements of modern law. However, the basic content did not change, as shown by the numerous examples analysed by Ōishi Kaichirō.[33] Formally, the contract of employment was drafted as a bilateral agreement, based on the consent of the workers themselves. It also mentioned explicitly the duty of the employer to pay a salary to the worker. However, the precise amount of the salary continued to be left undetermined. It was calculated *ex post facto* "on the basis . . . of the results of the work." Practically, as we shall see, it was impossible to know the amount of the wage earned until the end of the work, owing to the existence of a peculiar piece-rate wage system (*tōkyū chingin sei*).

As for other aspects, the contract of employment remained virtually unchanged in substance. For example, the rationale of the contract continued to be the money advanced by the employer. The worker, moreover, was still prohibited from changing her employment during the period of validity of the contract, and in case of violation of one of its clauses she had to pay a heavy penalty. No mention was made of the possibility of breach of contract or the payment of penalties by the employer. There is, instead, enough evidence to show that the payment of penalties was actually demanded from workers or their families. For example, in October 1906, a judge in the Court of Ueda (Nagano Prefecture) upheld the plea of a silk-reeler regarding the payment of a penalty of 60 yen for breach of contract.[34]

Only from the 1920s did the duties of the employer, to a certain extent, become more clearly indicated in employment contracts. This development reflected the emergence of unions (*jokō kyōkyū kumiai*) at the local level for the protection of female workers. However, the efforts of these unions, which were more often controlled by village officials and authorities than by the workers themselves, could not essentially change employer–employee relationships. Thus the workers continued to be practically bound to their employers by virtue of their debts.

The existence of this pre-modern employment relationship, together with the composition of the workforce, enables us to understand how the peculiar organization of labour and harsh working conditions could be introduced and maintained in Japanese silk filatures, as discussed below.

III. Labour Management

1. Working Conditions

Although working conditions could differ depending on the size of the enterprise, in the larger filatures set up after the middle of the 1880s they tended to be quite similar, as pointed out in many government reports. Apart from the case of small silk filatures located within the premises of the silk-reeler's household, the typical filature consisted of a central two-storied building housing the offices, the dormitories, a dining room, and other facilities for the workers; to this building were connected in rows, like the teeth of a comb, the mills for silk-reeling, re-reeling, cocoon-boiling, and so forth. Scattered around this central structure could be found the warehouses, the boiler plant and other buildings. The layout of the factory thus resembled that of a contemporary European silk filature. However, the fact that all the facilities for the workers' everyday lives were placed within the factory compound was distinctive of the Japanese filatures. All the factories were built of wood, in the local style. Private entrepreneurs could not afford costly structures of red brick like those of the Tomioka Silk Filature, which was built by the government. In general, the reeling departments were narrow, oblong buildings less than 7 metres wide, where the reeling machines were placed in rows on both sides facing the windows.[35]

As a whole, the workplace was designed with very little attention to the needs of the workers. The reeling machines were very low and the girls operating them had to sit at an angle, with the lower half of their bodies twisted at 30 degrees, because there was no space for their legs under the machine.[36] According to various reports, the light inside the mills was insufficient because the buildings were close together. The problem of poor illumination was particularly acute in the evening hours. According to an official survey of 1920, out of 603 silk filatures in Nagano Prefecture, only 36 could provide each worker with more than 6 candle-power for illumination, although the officials estimated that at least 10 candle-power was necessary for the delicate task of silk-reeling.[37]

The high degree of humidity in silk filatures was also detrimental to the health of the workers. Ishiwara Osamu, a sympathetic doctor who did a detailed study of the conditions of female textile workers, wrote: "In silk filatures, thick vapour always fills the place . . . It is quite strange that young girls who are not yet fully grown can maintain good health under such conditions."[38] In fact, tuberculosis and diseases of the digestive and reproductive organs were very common among female silk-reeling workers. Many also suffered from eye ailments, rheumatism and eczema, from having their hands constantly in hot water. Working conditions were particularly severe in the last month of the working season, when it could be extremely cold in the highlands of Suwa.

Shokkō jijō reports that at the turn of the century most of the silk filatures were not equipped with any heating system whatsoever. "In winter, female workers in silk filatures were usually affected with chilblains and frostbite, and in the afternoon the lower half of their legs became violet."[39]

Only toward the end of the Meiji period (1911) was a simple heating system finally adopted, since (*sic*) "the high number of workers affected by frostbite influenced quite considerably the efficiency of operations."[40] The heating system consisted of small pipes placed at the girls' feet, through which flowed hot water or steam during working hours.

Accidents were also frequent, often due to the explosion of poorly built boilers. It thus appears that the low level of fixed capital investment—one of the major features of the industry's development—resulted in an extremely poor working environment. Moreover, efforts to economize on fixed capital investment kept labour productivity to a very low level, a condition that was offset by low wages, long working hours, and high work intensity.

As for working hours, in 1902 a silk-reeler from Suwa opposed to the Factory Law made the following statement in a local newspaper:

> At present, we work 15 hours a day. In summer, from 4 a.m. to 7 p.m., and in winter, from 6 a.m. to 9 p.m. For meals we take just two or three minutes . . . [swallowing the food rather than eating it]. For us silk-reelers these working hours are necessary . . . For us every moment is precious. This idea of one hour and thirty minutes (the rest period indicated in the 1902 draft of the Factory Act) is really the cause of serious anxiety. At the longest, I wish it would be limited at most to about one minute after each meal.[41]

In general, at the beginning of the century working hours could vary from a minimum of 12 to 15 hours a day, as shown in table 1.7, which is based on the information provided by the silk-reelers themselves. Independent observers, however, report that in reality the upper limit was often exceeded.

According to Sano Zensaku, a pioneer of the study of commerce in Japan who visited the Suwa area in 1894, female workers in silk filatures worked on average 15 or 16 hours per day.[42] *Shokko jijō* reports that the average

Table 1.7. Working Days and Working Hours (1900 survey)

Silk filature	Working days per year	Daily working hours
Tōyōkan	210	15
Ryūjōkan	235	13
Shinryōsha	191	13
Imaikan	200	12
Kyūseikan	185	14
Shin'eisha	220	12
Shichiyōseisha	230	12

Source: *Seishi kankei shorui, 8, Hirano-mura yakuba, Meiji 23–35* [Documents on Silk-reeling, 8, Hirano Village Hall, 1890–1902], document kept at the Municipal Museum of Sericulture in Okaya.

working hours in Suwa did not go below 15 hours per day, but in case of favourable market conditions "it often happened that working hours could be extended up to 18 hours a day."[43]

The enforcement of the Factory Act in 1916 limited the daily working hours for female workers and for young men under 15 years old to 12 hours, at least in the factories with 15 or more employees (Article 3). However, for certain sectors like silk-reeling, the law allowed for an extension of the working time by another two hours a day for a period of 15 years from its enforcement. Further, Article 8 stipulated that the factory owner could extend the working hours another hour for a certain period not exceeding 120 days a year, provided that he obtained beforehand the approval of the administrative authorities.

Thus the silk-reelers of Suwa could continue the practice of long working hours by skilfully exploiting all these exceptions. For example, it is reported that in 1917 in Nagano Prefecture, 598 factories (the highest number in Japan) applied for an extension of working hours.[44] A female worker employed in the same year in a silk filature of Okaya recalls that: "At that time the factory was worse than a jail. We used to reel from 5 a.m. until the evening. When we were late, until 8 or 9 p.m., we rested for about 10 minutes in the morning at 9 a.m. and in the afternoon at 3 p.m. Earlier, people used to work all day long even without that brief rest."[45]

Another characteristic of silk filatures in Suwa was the high intensity of work, referred to by many observers. Sano Zensaku, for example, wrote that, notwithstanding the long working hours, "everybody was working hard, and there was no sign of lassitude. Unwilling to waste a single moment, none of the workers sat down at ease for their meals. They rather gathered in a hurry into the factory, having barely finished eating."[46]

Another observer in 1917 describes the tension, the intensity of work in the silk filature owned by the silk-reelers of Suwa, as something "unparalleled."[47]

How was it possible that, despite the harsh working conditions we have just seen, workers could be so highly motivated and ready to sacrifice themselves for such miserable wages? In order to understand this point we shall examine in the following pages the peculiar wage system and the organization of labour, which, together with the general social conditions of the period, in our opinion determined the high intensity of the work.

2. Wage System and Control of Workers

(1) The *Tōkyū chingin sei*

The peculiar wage system in Japanese silk filatures of the pre-war period was one of wages by ranking (*tōkyū*).[48] As already mentioned, this system is based on piece-rate wages, but was peculiar because the wage of each worker was calculated by comparing the productivity of that worker with the "average productivity" of all the workers. As a consequence, it was possible that for the same performance a worker could receive an inferior wage if, in the meantime, the average performance had gone up. Besides, since the "average productivity" could not be known until the end of the working day, the system created an extremely competitive working environment and, as a consequence, the high intensity of work we have just mentioned. On the other hand, the silk-reelers themselves knew in advance the total

Table 1.8. Wage Differentials in Suwa Silk Filatures

	Number of workers				
Daily wage in sen ($\frac{1}{100}$ yen)	Yamajō, 1889	Yamani, 1897	Yamajō, 1899	Yamajō, 1909	Yamani, 1911
50 or more				1 (0.8)[a]	
40–49			6 (6.8)	14 (10.9)	10 (13.7)
30–39		5 (3.8)	10 (11.4)	32 (25.0)	20 (27.4)
20–29		49 (37.4)	35 (39.8)	53 (41.4)	22 (30.2)
15–19	8 (16.0)	42 (32.1)	25 (28.4)	13 (10.2)	14 (19.2)
10–14	34 (68)	30 (22.9)	9 (10.2)	12 (9.4)	6 (8.2)
Less than 10	8 (16.0)	5 (3.8)	3 (3.4)	3 (2.3)	1 (1.4)

a. Figures in parentheses are percentages.
Source: Ishii Kanji, *Nihon sanshigyōshi bunseki*, p. 304.

Table 1.9. Wages for Female Workers, 1894–1903 (in sen)

Year	Silk-reeling	Cotton-spinning	Agricultural day labourers
1894	13	9	11
1895	13	10	11
1896	15	12	13
1897	18	14	16
1898	20	15	18
1899	22	17	17
1900	20	18	19
1901	20	19	20
1902	20	20	19
1903	20	21	19

Source: Sumiya Mikio et al., *Nihon shihonshugi to rōdō mondai* [Japanese Capitalism and the Labour Problem] (Tokyo Daigaku Shuppankai, 1967), p. 101.

amount of wages to be paid, since they usually fixed (*ex ante*) wages related to a certain placement in the final ranking without regard to the amount of work done.

In this sense, the wage system was perfectly tailored so as to foster a high intensity of labour, while keeping down its overall cost. Moreover, since the silk-reeling operation depended heavily on the individual skill of the worker, the system resulted in conspicuous wage differentials. As shown in table 1.8, at the end of the Meiji period a skilled worker could earn as much as four to five times more than an unskilled operator. Besides, the wage differential tended to grow during the period, since the system for computing wages was made more complex, in particular by the introduction of a system of fines based on qualitative standards (fineness, shine, existence of knots, and so on.) from the second half of the 1880s.

In general, wages on average were very low, just above the level of female day-labourers in agriculture (table 1.9). The table also shows that, from 1903 onward, the average wage for female workers in silk-reeling was inferior to wages in cotton-spinning, a more highly mechanized sector which employed large numbers of female workers. Basically, wages and working conditions in silk-reeling were thus related to the conditions prevailing in the rural sector, as could be expected considering the composition of the workforce. On the other hand, the higher wages paid to a small percentage of the workers could serve as a decoy in hiring new workers, or as an incentive for raising the level of work efficiency.

(2) The Role of the Supervisor

Besides the peculiar wage system, a strict system of controls on worker activities, during and outside of working hours, contributed to maintaining the high intensity of work in the filatures. In particular, the supervisor system—called the *kenban seido*—played a major role in ensuring the discipline of the workforce. According to the detailed, informative report on the *kenban seido* published in 1917 in *Shakai seisaku jihō*, in Suwa the supervisor (*kenban*) was a male employee who controlled the activities of a group of female workers (20 to 50 workers depending on the size of the factory).[49] Although in other regions, especially in the Kansai area, supervisors might be chosen among older female workers, in the silk filatures of Suwa this task was always assigned to male employees. Perhaps it was assumed that female supervisors could not maintain strict discipline under the harsher working conditions prevailing in Suwa.

The *kenban*, as already mentioned, was also in charge of recruiting each year the girls who were to work under his surveillance during the reeling season. Usually, as pointed out by Takizawa Hideki, the supervisor concentrated his efforts in a limited, specific area of the country.[50] This practice, besides economizing on hiring costs, had the advantage of reducing the risk of in-group conflicts and friction while stressing, on the other hand, competition among working groups from different regions. Moreover, it could be functional in maintaining work morale by preventing some girls from feeling isolated in the factory surroundings. It is interesting to note that the Japanese Imperial Army followed the same system of regional recruiting for its conscription system.

As for the supervisor's tasks, besides controlling the operations of the workforce under his surveillance, every morning the *kenban* read to the workers the production results achieved by each worker during the previous day. The report also mentions that every 15 days, when the production results of all the departments were compared, the supervisor received a red and white flag if his department had achieved outstanding results, or a black flag for poor results. The flag was paraded through the department either to keep up the workers' pride if they had been successful, or, if they had not, to encourage them to work harder in order to compete with their colleagues in other departments.

Since the wage and the promotion chances of the supervisor were directly related to the production results in his department, "the *kenban* were eager to use any methods whatsoever so as to improve work efficiency." This fact often led to the authoritarian, despotic use of their power that is often mentioned in the literature of the period.[51]

(3) The Dormitories

The system of dormitories for female workers was widespread in the Japanese textile industry, especially in Suwa's silk filatures. The system emerged partly because of the need to accommodate girls from distant districts. From the 1890s, moreover, they were relied upon to keep workers from leaving the factory before the expiration of their contracts, as well as to prevent their outright kidnapping by competitors.

Dormitories also made it possible to extend the supervision of workers beyond working hours, giving the factories control over their private lives as well. As a silk-reeler candidly declared in 1894: "Although the accommodation of female workers in dormitories requires a certain amount of expense, it makes possible full control of their conduct, from the time they wake, to the commencement of work, to the time they go to bed."[52]

The "Dormitory Regulations" printed in 1904 by a major silk filature in Suwa help us to imagine how the workers were treated in these institutions.[53] First, the regulations prescribe that all the workers, without exception, are to live in the dormitory. They were forbidden from commuting from boarding houses or even from their own homes. Since a similar rule can be found in the regulations of other silk filatures (for example, in those of the Ozawa Group),[54] it is likely that it was a generally applied rule in the Suwa area. This practice only too clearly shows that the dormitory was not simply an institution for accommodating workers from distant areas, but also had a specific function related to the control of the workforce.

From the regulations we can also infer that a strict, almost military discipline was enforced in the dormitories. For example, older women controlled the discipline, "patrolling inside and outside the rooms, . . . constantly observing the behaviour of the workers. In case of misconduct by some of the workers, they report it immediately to the director and solicit punishment for the workers concerned."[55]

Permission to leave the factory was also severely restricted. *Shokkō jijō* reports that female workers were generally forbidden to leave the factory grounds in the silk filatures of Suwa. Only in case of dire necessity could they leave, and then had to be accompanied by a supervisor or some kind of attendant.[56] Some factories kept a register of worker outings, in which was indicated not only the date and time of the outing, but also its destination and purpose.[57]

Concerning hygienic standards for the dormitories, a wealth of information is provided by a survey of the Ministry of Agriculture and Commerce conducted in 1917 and covering 120 silk filatures in Nagano Prefecture.[58] Dormitories were wooden structures, usually with few and small windows so as to keep out the cold. In some cases, however, there were no windows at all. The dormitories were usually located on the second floor of the main building, just above the offices—in other words, directly under the control of the management. Usually they were partitioned into big rooms, one of which, in one case, was more than 190 square metres in size. The average space set aside for each worker, however, was less than one mat (generally less than two square metres), and in some cases, when two or more girls had to share bedding, even less than half a mat was available (table 1.10).

The wooden floor was usually covered in the Japanese style with *tatami* mats (thick straw matting), or, in some cases, with thinner straw mats (*waramushiro*). Bedding was placed on these mats, and often shared by two or even three girls at the same time. Of 50 large silk filatures surveyed, only

Table 1.10. Mat Area Available for Each Worker in Silk Filature Dormitories in Nagano Prefecture, 1917

Number of mats	Number of silk filatures
Less than 0.5	7
0.51–0.6	12
0.61–0.7	23
0.71–0.8	23
0.81–0.9	22
0.91–1.0	13
More than 1.0	16
Total	116

Source: Nōshōmushō [Ministry of Agriculture and Commerce], *Kōjō kantoku nenpō, Taishō 5 nen* [Annual of Factory Inspection, 1916] (1918), p. 101.

four provided a set of bedding for each worker. Needless to say, in these conditions it was impossible to get enough rest after long hours of work. The unhealthy life of the dormitories greatly contributed to the diffusion of infectious diseases, particularly tuberculosis, among the workers.

3. "Welfare" Facilities

We shall now consider the function and realities of so-called "welfare facilities" in Japanese silk filatures. Medical, educational, and recreational facilities, which have become a standard part of employee benefits today, were only introduced in the major silk filatures in Suwa toward the end of the Meiji period (1912). This development reflected the growth of the industry and, in particular, its consolidation after the turn of the century. It seems however, that their introduction, as in other industrial sectors, was related, in the first place, to the difficulties of acquiring a stable workforce.

As a matter of fact, from the second half of the 1880s female worker mobility in silk filatures was rising, notwithstanding the binding clauses of the employment contract, the strict control on worker activities, and even the introduction of the complex system of worker registration by the Association of Silk-reelers. Female workers, taking advantage of the competition among silk-reelers, increasingly refused to renew their contracts at the end of the reeling season, or simply escaped from the factory. In this way they made their protest against harsh working conditions.

The introduction of "welfare facilities" was probably conceived as a countermeasure against this form of worker resistance. They were designed first to make factory life more attractive, thereby increasing the chances of recruiting and keeping new workers. Second, as we shall see, they functioned to extend further the sphere of control over workers.

Medical care, although referred to as part of "welfare facilities" (*fukuri shisetsu*), had little connection to the improvement of the "welfare" of workers. In the best of cases it was simply an attempt to provide a belated remedy for the damage caused by harsh working conditions. It is more likely, however, that the introduction of medical facilities was related to the need to prevent workers from leaving the factory.

In general, medical facilities were introduced from the early phases of the industry's development. An official report mentioned the existence of infirmaries in some Suwa filatures already in the 1890s.[59] Generally, however, they were no more than simply wooden barracks for recuperating patients. For example, at the Sanzensha, a large factory employing more than 800 workers, the infirmary built in 1894 was a small building of 60 square metres for the convalescence of up to 36 patients. According to the same report, some companies, including the Sanzensha, also paid for medicine and meals for the workers who suffered injuries due to occupational causes. In all other cases, only meals were provided.[60]

Medical facilities and various forms of assistance became more widespread after the Russo-Japanese War (1904–1905). Then, in 1910, the silk reelers of Suwa inaugurated their own hospital (Hirano Seishi Kyōdō Byōin).[61] In the meantime, especially after the turn of the century, major silk filatures had begun to organize educational activities for their workers. A 1901 report points out that already at that time some of the largest factories organized lectures for female workers on a regular basis.[62] In the silk filatures of the Katakura Group, "regular lectures were held from one to three times every month, and special lectures were also held from time to time . . . The lectures dealt with themes such as morals, housekeeping, hygiene, technology, and so on. These were given by Buddhist priests, teachers or members of the staff." In one factory, the Shinkōkan, even policemen were invited to give lectures to the workers.[63]

Educational activities became more systematic toward the end of the Meiji period, probably as part of the national campaign for "popular education" (*tsūzoku kyōiku*) launched after the Russo-Japanese War. In 1911, at the Ozawa Group, "instructive, edifying lectures on education, morals, housekeeping, hygiene, and so on were given three or four times a month for a couple of hours after the end of work. Usually the lectures were given by the owner or the supervisors, although Protestant pastors, Buddhist priests, and specialists could also be invited."[64]

We shall now examine in some detail a booklet, *Kōjōkun* [Ethics for Female Workers], which was published in 1910 and presumably used in the silk filatures of Suwa for ethics lectures.[65] According to *Okaya-shi shi* [History of the City of Okaya], this booklet was recommended by the trade association of Suwa silk-reelers. There are also records showing that Katakura and other major silk filatures ordered several copies of the booklet,[66] and it can be assumed to be a fairly typical example of an ordinary text used for the moral education of female workers.

The booklet is written in simple language easily understood by young girls. All the ideographs have syllabary *kana* at the side to facilitate reading. The text emphasizes first the contribution and importance of silk exports for the economic and military development of Japan. Developing his own version of the official slogan *fukoku kyōhei* ("enrich the country and strengthen its arms"), the author sees the mission of female workers as identical to that of a soldier. To quote the author's words:

> Japan has emerged with one bound as a first-class nation after the Sino-Japanese and the Russo-Japanese wars. However, no matter how high might be its position in the world, Japan cannot achieve the final victory only by relying on force. Together with (military) might, a country must possess wealth. . . . Each of you should work as hard as you can in order to increase the wealth of the nation. You should strive so that Japan will not be defeated by foreign nations.[67]

By appealing to Japanese nationalism in this way, the author tries to implant in the workers' minds a high sense of responsibility and pride in their occupation. For the same reason, he also underlines the fact that the imperial family had always encouraged sericulture, and that the Empress was actually practising it in the Imperial Palace in Tokyo.

Further, the booklet emphasizes, together with the Confucian virtue of filial piety, the virtue of loyalty (*chūgi*) toward the master. The emphasis on filial piety, a deeply felt sentiment in Japan, might have been quite common in popular texts, but the emphasis on loyalty, regarded as the highest virtue of the warrior, was rather out of place in the context of the lives of young girls of the Meiji period. However, it is obvious that the usual importance given to loyalty reflected the author's tacit condemnation of worker mobility as anti-social. On this point the author writes:

> Not only could the warriors of former days not serve two masters. . . . You, too, must be determined not to have two masters. . . . There are workers who move daily from filature to filature, selling themselves for a small gain, without showing any gratitude to their masters. These workers deserve to be called vagabonds. Once you choose a master, you should strive to serve him for all the years to come.[68]

We can infer from the content of this booklet that lectures on morals in the filatures were regarded as a means of keeping workers docile and obedient. Of course, it is hard to believe that this kind of lecture could have a strong impact on the minds of the workers. After many hours of hard work, they could not have been that eager to listen. Besides, the hardships of their lives were certainly more real and impressive than the shallow rhetoric of these booklets.

The introduction of so-called "welfare facilities" looks like an attempt to strengthen the control of the workforce in face of growing resistance. On the other hand, the reality of factory life continued to be dominated by long working hours, high work intensity, and almost limitless control over workers' lives—factors which, in our opinion, played a determining role in the industry's growth in pre-war Japan.

Notes

1. Mitsuya Tetsuya, *Seishi-gaku* [The Study of Silk-reeling], vol. 2, p. 5, as cited by Okumura Shōji, *Koban, kiito, watetsu* [Gold Coins, Raw Silk, Japanese Steel] (Iwanami Shoten, Tokyo, 1973), p. 108.
2. Sanshi Dōgyō Kumiai Chūōkai [Silk-reelers Industry Union Central Committee], *Kiito seisanhi ni kansuru kenkyū* [Study on Raw Silk Production Costs], in Okumura, *Koban, kiito, watetsu*, p. 108.
3. On the production process of raw silk in Japan, see Katō Sōichi, *Nihon seishi gijutsu-shi* [History of Silk-reeling Technology in Japan] (Seishi Gijutsushi Kenkyūkai, 1976), and Okumura, *Koban, kiito, watetsu.*
4. Okumura, *Koban, kiito, watetsu*, p. 81.
5. See *Shinao sanshigyo-shi* [History of the Silk Industry in Shinano], vol. 2 (1937).
6. Shibusawa Eiichi, *Shibusawa Eiichi denki shiryō* [Documents on the Life of Shibusawa Eiichi] (Shibusawa Eiichi Denki Kankōkai), vol. 2.
7. Cited in Katō, *Nihon seishi gijutsu-shi*, p. 89.
8. Nagaoka Shinkichi, *Sangyō kakumei* [The Industrial Revolution] (Kyōikusha Rekishi Shinsho, 1979), p. 102.
9. Okumura, *Koban, kiito, watetsu*, pp. 106–107.
10. Katō, *Nihon seishi gijutsu-shi*, p. 116.
11. Wada Hide, *Tomioka nikki* [Tomioka diary] (Chuō Kōronsha, 1978), pp. 21–22.
12. Katō, *Nihon seishi gijutsu-shi*, p. 116.
13. Hirano-son Yakuba [Hirano Town Hall], *Hiranoson-shi* [History of the Town of Hirano], vol. 2 (1932), p. 340.
14. Ishii Kanji, "Sangyō shihon (2): Kengyō" [Industrial Capital (2): The Silk Industry], in Oishi Kaichirō, ed., *Nihon sangyō kakumei no kenkyū* [Studies on the Industrial Revolution in Japan], vol. I (Tokyo Daigaku Shuppankai, Tokyo, 1975), p. 175.
15. *Tokyo asahi shimbun*, 6 January 1931. Cited in Yamada Moritaro, *Nihon shihon-shugi bunseki* [Analyses of Japanese Capitalism] (Iwanami Shoten, Tokyo, 1934), p. 41.
16. Cited in Shinohara Akira, "Seishi kikai no rekishi—Meiji shoki made no hatten katei" [The History of Silk-reeling Machines: Development up to the Early Meiji period], *Journal of the Faculty of Textile Science and Technology, Shinshū University*, no. 74, ser. B, Engineering, no. 13, December 1978: 28–29.
17. Ibid.
18. Ishii Kanji, *Nihon sanshigyo-shi bunseki* [Analysis of the History of Japan's Silk Industry] (Tokyo Daigaku Shuppankai, Tokyo, 1972), p. 245.
19. Ibid.
20. See Minorikawa Naosaburō, *Minorikawa Naosaburō-Ō jiden* [Autobiography of Minorikawa Naosaburō) (Minorikawa Naosaburō-Ō Kankokai, Tokyo, 1933).
21. Nōshōmushō Nōmukyoku [Ministry of Agriculture and Commerce, Agricultural and Commerce Bureau], *Zenkoku seishi kōjō chōsahyō* [Survey of Silk-reeling Factories throughout Japan], no. 6 (1913), p. 171.

22. Nōshōmushō Shōkōkyoku [Ministry of Agriculture and Commerce, Commerce and Industry Bureau], *Shokkō jijō* [Conditions of Workers], vol. 1, 1903 (1976), p. 163.
23. See Andō Yoshio, ed., *Kindai Nihon keizaishi yōran* [A Survey of Japan's Modern Economic History] (Tokyo Daigaku Shuppankai, Tokyo, 1975), p. 98.
24. *Hiranoson-shi*, vol. 2, p. 157.
25. The decrease was particularly evident in Kami-ina district, probably reflecting the growth in this region of cooperatives of silk-reelers that were competing with Suwa's capitalist enterprises. On the development of cooperatives in Kami-ina, see Oshima Eiko, "Issen Kyūhyaku Nijūnendai ni okeru kumiai seishi chitai ni okeru nōgyō kōzō to tei-rishikin no igi" [Silk-reeling Cooperatives' Production of High-quality Silk in the 1920s: The Significance of Low-interest Funds and Agricultural Structure in the Silk-reeling Cooperatives' District of Kam-ina in Nagano Prefecture], *Rekishigaku kenkyū* (November 1980).
26. Ishii, *Nihon sanshigyo-shi bunseki*, p. 265.
27. Ibid., and Nakamura Masanori, "Seishigyo no tenkai to jinushi-sei" [The Development of Silk-reeling and the Landlord System], *Shakai keizaishigaku*, vol. 32, nos. 5–6 (February 1967).
28. On this point see the diary of a landlord's administrator cited in Nakamura Masanori, "Seishigyō no tendai to jinushi-sei," p. 60.
29. On the definition of "*dekasegi*-type" of workers, see Okochi Kazuo, *Reimeiki no Nihon rōdō undō* [Japanese Labour Movement in the Early Period] (Iwanami Shoten, Tokyo, 1953), pp. 4–10.
30. On factory dormitories see section 2(3). As for the system of worker registration introduced as a measure against job mobility, see Ishii, *Nihon sanshigyo shi bunseki*, pp. 277–290, and Tōjō Yukihiko, "Seishi Dōmei no joko toroku seido no hensen ni tsuite" [On Changes in the Registration System of Female Workers under the League of Silk Reelers], *Tochi seido shigaku*, No. 101 (October 1983). Questioning the result of previous research, Tōjō underlines the fact that the system was rather ineffective in reducing worker mobility.
31. Commerce and Industry Bureau, *Shokko jijō*, pp. 178–179.
32. The original document is in the Okaya Municipal Museum of Sericulture (Shiritsu Okaya Sanshi Hakubutsukan).
33. Oishi Kaichirō, "Kōjō keiyakusho no hensen kara mita seishigyō chinrōdō no keitai henka" [The Transformation of Wage Labourers in Silk-reeling Seen through the Change of Employment Contracts," *Shakai kagaku kenkyū*, vol. 24, no. 2 (1972): 87.
34. *Nagano kenshi, Kindai Shiryōhen, sanshigyō* [History of Nagano Prefecture, Documents on the Modern Period, Silk Industry], vol. 5 (3) (1980), pp. 874–875.
35. Maps, plans, and photographs of silk filatures are collected at the Okaya Municipal of Sericulture. On the structure of silk filatures around Suwa, see also *Hiranoson-shi*, vol. 2, p. 313, and Ministry of Agriculture and Commerce, *Kōjō kantoku nenpō, Taishō 5 nen* [Annual of Factory Inspection, 1916] (1918), p. 91.
36. Ibid., p. 92
37. Tokyo Chihō Shokugyō Shōkai Jimukyoku, *Kannai seishi kōjō chōsa* [Survey of Silk Filatures within the Jurisdiction of the Employment Office for the Tokyo Region] (1925), p. 180. See also Mitani Tetsu, "Naganoken seishigō ippan" [Outline of Silk-reeling in Nagano Prefecture], *Dai Nippon sanshi kaihō* [Reports on the Silk-reeling Industry of Japan], no. 149 (1904).
38. Ishiwara Osamu, "Jokō to kekkaku" [Female Workers and Tuberculosis], in *Seikatsu kōten sōcho*, vol. 5, 1970, p. 179.

39. Commerce and Industry Bureau, *Shokkō jijō*, vol. 1, p. 189.
40. Ministry of Agriculture and Commerce, *Kōjō kantoku nenpō, Taisho 5 nen*, p. 91.
41. *Shinano mainichi shimbun*, 17–18 February 1902. Cited in Ishii, *Nihon sanshigyō shi bunseki*, pp. 363–64.
42. Sano Zensaku, Kobayashi Wakai, *Yamanashi-en Ichien, Nagano-ken Suwa/Ina shisatsu hōkokusho* [Report on a Survey of Yamanashi Prefecture and the Suwa and Ina districts of Nagano Prefecture] (1894) (Hitotsubashi University Library).
43. Commerce and Industry Burearu, *Shokkō jijō*, vol. 1, p. 174.
44. Ministry of Agriculture and Commerce, *Kōjō kantoku nenpō, Taisho 6 nen*, [Annual of Factory Inspection, 1917] (1919), p. 97.
45. Chiiki Shakai Kenkyūjo, Kōnenreisō Kenkyū Iinkai, *Kōnenrei o okiru, 14, Ken nōgyōka no otoshiyori-tachi Nagano-ken Suwa-shi kōnan no kurashi* [Living to Advanced Age (14), The Lives of Older Farming People of the Town of Suwa, Nagano Prefecture] (1981), p. 56.
46. Sano and Kobayashi, *Yamanashi-ken Ichien, Nagano-ken Suwa/Ina shisatsu hōkokusho*.
47. Fujii Tei, "Seishi jokō to suettingu shisutemu" [Female Workers in Silk-reeling and the Sweatshop System], *Shakai seisaku jijō*, no. 10 (June 1920): 21.
48. On the *Tōkyū chingin sei, see* Ōishi Kaichirō, "Nihon seishigyō chinrōdō no kōzōteki tokushitsu—Tōkyū chingin sei o chūshin toshite" [Structural Characteristics of Wage Labour in Japanese Silk-reeling: Centring on the Ranked Wage System], in Kawashima Takemi et al., *Kokumin keizai no shoruikei* [Aspects of the National Economy] (Iwanami Shoten, Tokyo, 1968), and Ishii, *Nihon sanshigyō-shi bunseki*.
49. Fujii Tei, "Seishi jokō to suettingu shisutemu." Unless otherwise noted, our description of the *kenban seido* is based on this report.
50. Takizawa Hideki, *Nihon shihonshugi to sanshigyō* [Japanese Capitalism and the Silk Industry] (Miraisha, Tokyo, 1978), pp. 395–408.
51. See, for example, the revealing book of Sakura Takuji, *Seishijokō gyakutaishi* [The History of the Cruel Treatment of Female Workers in Silk-reeling] (Tokyo, 1927).
52. Sano and Kobayashi, *Yamanashi-ken Ichien, Nagano-ken Suwa/Ina shisatsu hōkokusho*.
53. *Nagano kenshi, Kindai Shiryōhen, sanshigyō*, vol. 5 (3), pp. 869–870.
54. Tokyo Kōtōshōgyō Gakkō [Tokyo Higher Commercial School], *Shokkō toriatsukai ni kansuru chōsa* [Survey on the Treatment of Workers] (1911), p. 89.
55. *Nagano kenshi, Kindai Shiryōhen, sanshigyō*, vol. 5 (3), pp. 869– 870.
56. Commerce and Industry Bureau, *Shokkō jijō*, p. 203.
57. One of these registers, compiled in 1910, is now reprinted in *Nagano kenshi, Kindai Shiryōhen, sanshigyō*, vol. 5 (3), pp. 878–879.
58. Ministry of Agriculture and Commerce, *Kōjō kantoku nenpō, Taisho 5 nen*, pp. 91–131.
59. See Ministry of Agriculture and Commerce, Commerce and Industry Bureau, *Kakukōjō ni okeru shokkō kyūsai sono ta jikeiteki shisetsu ni kansuru chōsa gaigyō* [Summary of a Survey on Worker Relief and Other Assistance Facilities in Factories] (1903).
60. Ibid., p. 37.
61. Ministry of Agriculture and Commerce, *Kōjō kantoku nenpō, Taisho 5 nen*, p. 124.
62. Commerce and Industry Bureau, *Kakukōjō ni okeru shokkō kyūsai sono ta*

jikeiteki shisetsu ni kansuru chōsa gaiyō, pp. 5–6.
63. Ibid., pp. 17–18.
64. Tokyo Kōtōshōgyō Gakkō, *Shokkō toriatsukai ni kansuru chōsa*, pp. 15–16.
65. Katō Tomomasa, *Kōjō kun* [Precepts for Female Workers] (1910).
66. *Okayashi-shi* [History of the Town of Okaya], vol. 2, p. 590.
67. Katō Tomomasa, *Kōjō kun*, pp. 4–6.
68. Ibid., pp. 22–23.

Chapter——2

The Coal-mining Industry

Yutaka Nishinarita

In pre-war Japan, coal-mining, along with textile manufacturing, was among the leading industries employing women. The proportion of women working in these two industries differed greatly: in mining, the overwhelming majority of workers were men, while in textiles women far outnumbered men. The working conditions of women in the textile industries have been the subject of much research, while this aspect of coal-mining has received little attention, for two major reasons. First, the mining communities were located in geographically isolated areas, they were hostile to outsiders, and their women were reclusive. Second, there is a lack of documents and other materials upon which to base research. Such publications as *Shokkō jijō* [Conditions of Workers], edited by the Ministry of Agriculture and Commerce in 1903, provide an in-depth understanding of the labour conditions of female textile workers. There are no such materials on coal-mining, and statistics on women workers in nationwide industries are available only from 1914. This chapter examines the rise and decline of the female labour force in coal-mining against the backdrop of development and change in mining technology.

Several considerations should be kept in mind in analysing the data in this study. First, we must pay attention to regional differences in mining techniques and in the working conditions of women, stemming from the size of the coalfields and the scale of the mines. Second, because the study of mining technology and female workers falls in the category of labour management, such issues as the level of technical skills, the adoption of new technology by the management, and the relationship of these factors to the rise and eventual decline of women workers must be addressed. Third, because the introduction of new technology generally brings about conflict between labour and management, whether the new technology is adopted successfully or not largely depends on labour–management relations. There are numerous examples in modern history of workers obstructing the adoption of new technology. The mechanization of the coal-mining industry was relatively smooth, and although it resulted in the eventual exclusion of women from its workforce, it is important to understand what conditions made the smooth transition possible.

I. Characteristics of Female Mine Labour

1. Development of the Labour Force

Evidence of women working in mines can be traced back to relatively old documents. For example, *Kōzan shiryōchō* [Survey Materials of Mines], published by the Industrial Department of Nagasaki Prefecture in 1884, states:

> Higashi Matsuura county, Kishiyama village, Aza Hase leased land, 1,500 *tsubo*.
> 1. Miners' wages: men, 25 sen; women, 18 sen.
> 2. Average number of workers per year: 700, of which 350 men, 350 women.
> 3. Coal extracted per work unit (a work unit consists of one man and one woman) 800 *kin* (480 kg) per month.[1]

From the above, we can surmise that women were already well incorporated into the two-person work teams. However, before the industrial revolution, mines were small in scale, and there were fewer women working in them. The majority of miners were part-time farmers who lived in adjacent areas and commuted to the mines or took on mine work during the slack season. Mining was not considered a full-time occupation, and it was not until after the industrial revolution, when mining no longer was a part-time job, that women became a significant part of the labour force in the mines.

The mechanization of coal transportation was a major development that marked the beginning of the industrial revolution in mining. Specifically, this meant the installation of a winch in the shafts. The use of the winch spread rapidly between 1890 and 1900, around the time of the Sino-Japanese War. Notably, however, mechanization in coal-mining did not go beyond the mechanization of the conveyance system in the main shafts. The coal-extracting process continued to be done by hand, with the use of simple tools.

Except when working on very thin seams, the most common digging method at the time was the pillar method. A main shaft was bored along the slant of the coal seam, and smaller shafts dug in checkerboard fashion at 10-*ken* (18-metre) intervals perpendicular to the main shaft. Three *ken* were used as the coal face, and seven as pillars to protect the shafts (fig. 2.1). The kind of labour involved in chipping out and extracting the coal was extremely isolated and dispersed. Each coal face was hewed with a pick, and

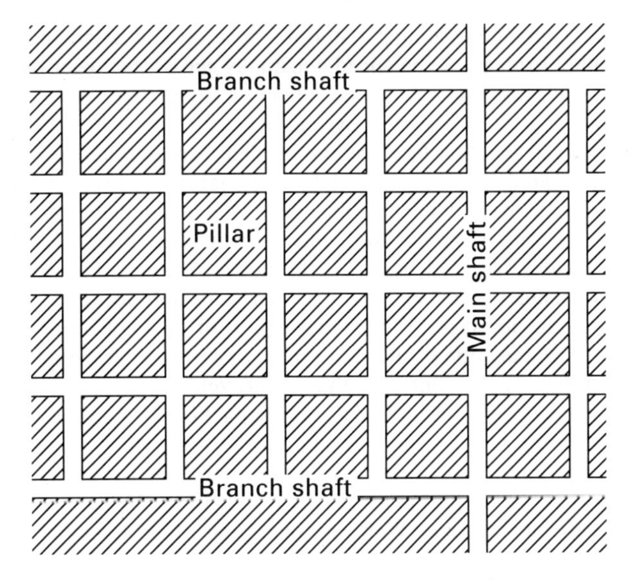

Fig. 2.1.

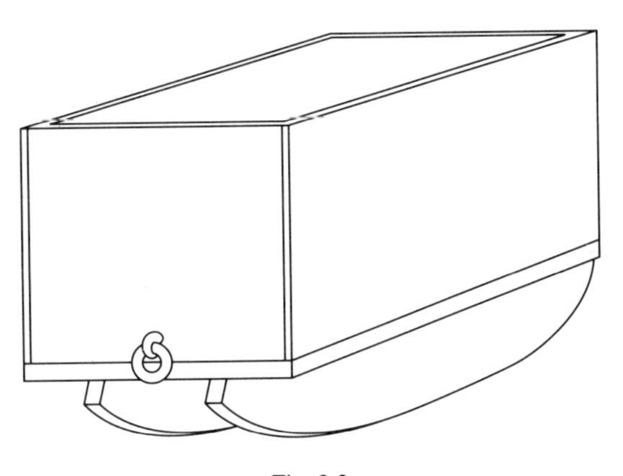

Fig. 2.2.

the pieces were carried by hand to a coal car. In the thicker seams, the coal thus extracted from the coal face was loaded into baskets holding between 50 and 100 *kin* (30 to 60 kg), and two of these baskets were balanced on a shoulder carrying pole, or buckets were hauled one at a time to the main shaft. In the thin seams, 100–200 *kin* (60–120 kg) of coal were extracted and loaded into bamboo baskets placed on sledges (fig. 2.2), which were pushed to the main shaft. One of the most demanding methods of carrying these baskets involved assuming a stooped position and balancing two baskets suspended from a pole balanced diagonally across the lower back.[2]

Ironically, because there were no advances in removal techniques, the more mechanized the conveyance system became, the greater the demands on the workers at the coal face. As the burden increased and working conditions deteriorated, more rigorous labour management became necessary, and the *naya* (stable) system came into being. *Naya* foreman recruited their own miners, settled them in bunkhouses, controlled their lives, allotted jobs, supervised work, and patrolled the mines. By making available various types of loans to workers, the foremen were able to keep the miners in virtual human bondage. The intensification of coal-collecting work demanded by mechanization of the conveyance system was perpetuated by the coercive supervision and control of the *naya* system.

The new conveyance system had a great impact on the composition of the workforce. Mining could no longer continue as a seasonal occupation sustained by part-time farmers; the new system demanded regular output. A stable workforce was required, and in the course of the industrial revolution following the Sino-Japanese War, the recruiting of professional, full-time miners, with preference for those with families, began. A 1906 recruitment announcement for Mitsui's Tagawa mine clearly illustrates this trend: "First preference to families consisting of a couple with a child of 12 or 13, all three of whom will be required to work; plus one elderly person to serve as cook."[3] At Mitsui's Miike mine, a 1905 recruitment announcement offered contracts mainly to couples or miners with families.[4] As a result of this recruitment policy, the percentage of miners with families rose: in 1903, at Meiji mine, 50.8 per cent of all miners and 60.2 per cent of coal collectors had families; in 1906 at Futase mine, the figure was 72.6 per cent of all miners; in 1909 at the main mines in Chikuhō, Kyushu, it was 65.3 per cent of all miners.[5]

There are two reasons why miners with families were preferred. First, the morale of the workers was higher.[6] Second, it was obviously more efficient, in view of the way the coal was extracted, to have the men digging and the women hauling. Thus it was that the industrial revolution brought women into the workforce and made the hiring of working couples and families the dominant recruitment pattern. The increase in women miners can be seen in tables 2.1 and 2.2.

The establishment of the *naya* system and the formation of the female mine workforce were developments that went hand in hand. (The Miike mine, where the *naya* system was abolished in 1908, was an exception, indicating that female workers and the *naya* system were not always compatible.) As said of the Chikuhō mine, the work contracts were essentially not between foreman and the individual miner but between the foreman and the miner and his family.[7]

Social relations in mining communities followed the patriarchal system, and women worked long and exhausting hours: besides the demands of

Table 2.1. Gender Composition of Miners at Mitsui Kōzan's Miike Mines, 1896–1991

Year	Male (A)	Female (B)	Total (C)	$\frac{B}{C}$ (%)
1896	4,734	83	4,817	1.7
1897	5,080	81	5,162	1.6
1898	4,936	79	5,015	1.6
1899	5,077	75	5,152	1.5
1900	4,520	572	5,092	11.2
1901	5,205	681	5,886	11.6
1902	6,018	871	6,889	12.6
1903	6,235	1,136	7,374[a]	15.4
1904	5,895	1,253	7,148	17.5
1905	6,098	1,423	7,521	18.9
1906	7,322	1,619	8,941	18.1
1907	7,871	1,679	9,550	17.6
1908	8,352	1,971	10,323	19.1
1909	9,985	2,159	12,144	17.8
1910	10,031	2,409	12,437	19.3
1911	10,460	2,626	13,086[b]	20.1

a. Total uncorrected.
b. Total corrected.
Source: *Mitsui Kōzan 50-nenshi kō* [A Fifty-year History of Mitsui Kōzan], vol. 16 (Labour), Mitsui Bunko Collection.

their jobs in the mines, they performed all the household chores. Yamamoto Sakubei writes of their plight:

> The life of the female coal-miner was appalling. Returning black and grimy from a day's work in the pits with their husbands, they immediately had to start preparing meals. In those days there were no nursery facilities, so infants were placed in the care of others. Mothers returned from the mines to nurse their babies. Men would return from their work to bathe and sit back and relax, displaying their tattoos and drinking sake. This was the accepted behaviour of these lowly people, and no man would be found helping with what were designated as women's tasks. If a woman so much as protested, she would be beaten.[8]

The heavy double workload on women, as well as the restrictions placed on their work in the mines (e.g. women were prohibited from entering the mines when menstruating), contributed to the decrease in working rates of women miners.[9]

Table 2.3 provides statistics on labourers at Namazuta mine between 1914

Table 2.2. Gender Composition of Miners at Mitsubishi's Shinnyū and Namazuta Mines, 1894–1912

Year	Shinnyū				Namazuta			
	Male (A)	Female (B)	Total (C)	B/C (%)	Male (A)	Female (B)	Total (C)	B/C (%)
1894	1,200	200	1,400	14.3	1,015	342	1,357	25.2
1906	3,250	1,227	4,477	27.4	1,262	544	1,806	30.1
1908	—	—	—	—	1,480	728	2,208	33.0
1912	3,937	1,572	5,509	28.5	—	—	—	—

Source: Hayama Saburō, "Shinnyū tankō" [Shinnyū Mine] and "Namazuta tankō," [Namazuta Mine], unpublished manuscripts.

Table 2.3. Working Ratios of Miners at Mitsubishi Namazuta Mine by Job, 1914–1917 (percentages)[a]

	1914		1916		1917	
Job	Male	Female	Male	Female	Male	Female
Digger	52.0	48.0	57.0	51.0	53.0	49.0
Bracer	79.0	—	75.0	—	78.0	—
Hauler	86.1	—	85.8	—	86.7	—
Dresser	—	81.9	—	82.0	—	76.5
Mechanics	94.0	—	93.5	—	96.5	—
Operators	93.9	—	94.1	—	91.3	—
Misc.	77.6	74.0	78.0	77.2	78.4	77.1

a. Figures for April–September of years given.
Source: Hayama Saburō, "Shinnyū tankō" [Shinnyū Mine] and "Namazuta tankō," [Namazuta Mine], unpublished manuscripts.

and 1917. The working rates of women in the pits were extremely low, compared to the surface workers who dressed the coal or performed miscellaneous chores. With fewer women haulers, the workload on male diggers generally increased, and this in turn restricted the management's dependence on the family recruitment system.

2. Regional Differences[10]

The proportion of women workers, the jobs they were assigned, and the work-unit formations were all dependent on the location of the coal—whether the seams were highly dispersed or concentrated—and the labour market of the region. Differences can be noted between one mine and another. A breakdown of mines and work positions at major mines is given in table 2.4. The three main jobs filled by women were pitman (hauler, sub-hauler), coal dresser, and bracer. Four mines (Joban, Chikuhō, Miike, and Karatsu) employed women in the pits (haulers, sub-haulers), while in three mines (Chikuhō, Miike, and Karatsu) they worked as bracers. Women miners were concentrated in the above four mines, but Miike differed from the other three in its work-unit formations. At Joban, Chikuhō, and Karatsu, the basic one digger/one hauler work unit was the rule, and the rate of working couples was high—38.3, 39.9, and 33.9 per cent respectively, as of the end of 1910. The working conditions were better at Miike because the seams were deep and the work units were made up of two diggers to two haulers, which did not necessarily conform to the family recruitment pattern; the rate of working couples at Miike was a low 12.5 per cent (fig. 2.3).

In contrast to the four mines mentioned above, the mines of Hokkaido and Nishisonoki (Nagasaki Prefecture; see figure 2.3) employed very few

Table 2.4. Miners at Major Mines by Job, 1906

		Hewers	Bracers	Haulers	Pit operators	Misc. (under-ground)	Total (under-ground)	Dressers	Haulers (surface)	Surface operators	Misc. (surface)	Total (surface)	Grand total
Hokkaido (6 mines)	Male	3,076	188	1,733	49	1,408	6,454	1,074	184	151	1,383	2,792	9,246 (86.0)
	Female	—	—	—	—	—	—	1,293	74	—	17	1,384	1,384 (12.9)
	Minor	—	—	—	—	—	—	29	—	—	93	122	122 (1.1)
	Total	3,076 (28.6)[a]	188 (1.7)	1,733 (16.1)	49 (0.5)	1,408 (13.1)	6,454 (60.0)	2,396 (22.3)	258 (2.4)	151 (1.4)	1,493 (13.9)	4,298 (40.0)	10,752 (100)
Jōban (7 mines)	Male	2,226	246	441	129	297	3,339	76	211	338	495	1,120	4,459 (73.0)
	Female	1,099	—	12	—	145	1,256	345	—	—	25	370	1,626 (26.6)
	Minor	—	—	—	—	13	13	10	—	2	—	12	25 (0.4)
	Total	3,325 (54.4)	246 (4.0)	453 (7.4)	129 (2.1)	455 (7.4)	4,608 (75.4)	341 (7.1)	211 (3.5)	340 (5.6)	520 (8.5)	1,502 (24.6)	6,110 (100)
Chikuhō (25 Mines)	Male	17,570	2,418	1,224	949	1,646	23,807	985	1,131	3,404	3,828	9,348	33,155 (73.2)
	Female	8,316	293	6	—	510	9,125	2,115	71	—	600	2,786	11,911 (26.3)
	Minor	115	—	—	—	1	116	72	—	8	8	88	204 (0.5)
	Total	26,001 (57.4)	2,711 (6.0)	1,230 (2.7)	949 (2.1)	2,157 (4.8)	33,048 (73.0)	3,172 (7.0)	1,202 (2.7)	3,412 (7.5)	4,436 (9.8)	12,222 (27.0)	45,270 (100)
Miike (1 mine)	Male	1,906	269	738	285	553	3,751	32	409	1,423	768	2,632	6,383 (77.5)
	Female	997	28	—	—	222	1,247	269	3	—	313	585	1,832 (22.3)
	Minor	3	—	—	—	—	3	—	—	11	2	13	16 (0.2)
	Total	2,906 (35.3)	297 (3.6)	738 (9.0)	285 (3.5)	775 (9.4)	5,001 (60.8)	301 (3.7)	412 (5.0)	1,434 (17.4)	1,083 (13.2)	3,230 (39.2)	8,231 (100)

Karatsu	Male	3,153	1,343	268	106	279	5,169	148	431	537	421	1,537	6,706 (68.7)
(6 mines)	Female	1,561	248	18	—	34	1,861	786	130	—	52	968	2,829 (29.0)
	Minor	144	12	—	—	4	160	55	—	7	4	66	226 (2.3)
	Total	4,858 (49.8)	1,603 (16.4)	306 (3.1)	106 (1.1)	317 (3.2)	7,190 (73.7)	989 (10.1)	561 (5.7)	544 (5.6)	477 (4.9)	2,571 (26.3)	9,761 (100)
Nishiso-		1,111	—	69	153	56	1,389	—	96	178	399	673	2,062 (92.7)
noki	Male	—	—	—	—	—	—	—	—	—	163	163	163 (7.3)
(1 mine)	Female	—	—	—	—	—	—	—	—	—	—	—	—
	Minor												
	Total	1,111 (49.9)	—	69 (3.1)	153 (6.9)	56 (2.5)	1,389 (62.4)	—	96 (4.3)	178 (8.0)	562 (25.3)	836 (37.6)	2,225 (100)

a. Figures in parentheses are percentages.

Source: Ogino Yoshihiro, "Nihon shihonshugi kakuritsuki ni okeru tankō rōshi kankei no 2 ruikei" [Two Patterns of Labour Management Relations in Mines during the Establishment of Japanese Capitalism], *Enerugii kenkyū nōto*, 10. With corrections. Original data from Ministry of Agriculture and Commerce, Bureau of Mining, *Kōfu taigū jirei* [Examples of Treatment of Miners] (1908).

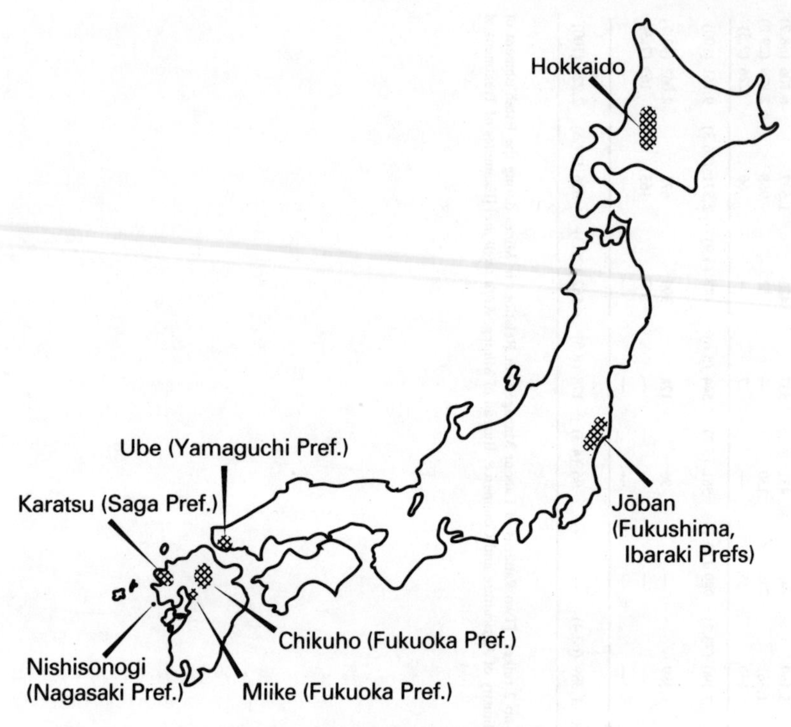

Fig. 2.3. Major Japanese coalfields.

women, and none worked underground. At the Hokkaido mines, women were surface workers, mostly coal dressers, and comprised 13 per cent of the total workforce. At Nishisonoki, the few women who were employed did miscellaneous surface chores. The labour market in these two areas posed difficulties for the recruitment of miners, particularly women. Nishisoniki was an island off Kyushu and its geographic location was a drawback. Hokkaido mines recruited from Tohoku, in northern Kyushu, where the farming class was undifferentiated. Thus it was that the single men who went as seasonal workers formed the majority of miners. Slope mining was employed, so that the hauling of coal did not require women labourers. Again in contrast to Chikuhō, Jōban, and Karatsu, the *naya* system was abolished in both Hokkaido and Nishisonoki during the early stages of the industrial revolution, and recruitment and management of the miners was done directly by the mine owners.

It can thus be seen that the jobs assigned to women and their percentage in the mines differed greatly by mine. Considering that Jōban, Chikuhō, Miike, and Karatsu mines produced 86 per cent of the nation's coal in 1906, and that Hokkaido produced only 10 per cent, and Nishisonoki even less, it is clear that women workers played a central role in the Japanese coal mining industry.

3. Characteristics of Women Miners: The 1920s

How do women miners compare with women in textiles, an industry that had a very high concentration of women before the Second World War? An examination of the nationwide statistics for the 1920s—which are very thorough—will shed light on the characteristics of women miners, their work, and and their lives.

(1) Age and Marital Status

The age and marital status of women miners and textile workers is given in table 2.5. The age range of the miners is older than that of textile workers: 71 per cent of textile workers were under 20, and the proportion of young women was overwhelmingly greater. Miners under 20 comprised only 24 per cent of the total, and 58 per cent were over 25. This contrast in age composition shows that young women made up the core of the textile workforce, while in the mines it was the middle-aged women. This contrast is also seen in the marital status of the women in these two industries: only 12 per cent of the textile workers, as opposed to 75 per cent of the miners, were married.

A closer look at the figures reveals that the age composition of women working in the three underground jobs (digger, hauler, supporter) differed from that of the coal dressers, who were surface workers. A large proportion (36 per cent) of the coal dressers were young girls under 20, but the figure is far lower than that for women over 25 working underground. As might be expected, more of the women working underground were married: 80 per cent, as opposed to 58 per cent of the coal dressers. Thus it can be concluded that the jobs underground were generally carried out by working couples (the family recruitment pattern), and that the proportion of young, unmarried coal dressers was high.

(2) Length of Employment

The average number of years women worked as miners is given in table 2.6. In the case of textile workers, half worked for less than three years, indicating a very low professional rate. In contrast, over half the miners were employed for over five years, indicating a high rate of professionalism. As with age composition and marital status, there were differences between the underground and surface workers. Those who worked over five years underground were close to 60 per cent (70 per cent in the case of supporters), but only 37 per cent of the surface workers worked for that length of time; 45 per cent of them left their jobs within three years. This difference in years of employment can be explained by the relatively high level of skills that the underground workers had to master; coal dressing was a simple task that required no training.

Table 2.5. Age and Marital Status of Women Miners, 1924

Age	Mining industry					Women textile workers
	Digger	Hauler	Dresser	Bracer	All	
<15	438 (6.1)[a]	1,917 (6.3)	1,728 (14.0)	242 (4.5)	4,880 (7.5)	211,002 (35.7)
16–19	1,152 (16.0)	4,915 (16.3)	2,725 (22.0)	546 (10.2)	10,586 (16.3)	206,715 (35.0)
20–24	1,476 (20.5)	5,834 (19.3)	1,996 (16.1)	864 (16.1)	11,530 (17.7)	99,332 (16.8)
25–29	1,303 (18.1)	5,383 (17.8)	1,355 (11.0)	1,090 (20.4)	10,463 (16.1)	29,149 (4.9)
30–39	1,850 (25.7)	7,876 (26.1)	2,105 (17.0)	1,571 (29.3)	16,023 (24.7)	24,371 (4.1)
40–49	888 (12.3)	3,922 (13.0)	1,870 (15.1)	952 (17.8)	9,761 (15.0)	14,010 (2.4)
>50	91 (1.3)	352 (1.2)	591 (4.8)	90 (1.7)	1,717 (2.6)	6,596 (1.1)
Total	7,198 (100.0)	30,199 (100.0)	12,370 (100.0)	5,355 (100.0)	64,960 (100.0)	591,175 (100.0)
Married	5,695 (79.1)	23,964 (79.4)	7,193 (58.1)	4,549 (84.9)	48,655 (74.9)	71,123 (12.0)
Single	1,503 (20.9)	6,235 (20.6)	5,177 (41.9)	806 (15.1)	16,305 (25.1)	520,052 (88.0)

a. Figures in parentheses are percentages.

Source: Cabinet Statistics Bureau, *Rōdō tōkei jitchi chōsa hōkoku* [Report of a Survey on Labour Statistics] (1924).

Table 2.6. Length of Employment of Women Workers, 1924

Age	Coal-mining industry					
	Digger	Hauler	Dresser	Bracer	All	Women textile workers
<1	522 (7.3)[a]	2,068 (6.9)	2,097 (17.0)	262 (4.9)	6,210 (9.6)	126,370 (21.4)
1–3	1,341 (18.7)	5,173 (17.1)	3,533 (28.6)	648 (12.1)	12,928 (19.9)	193,548 (32.8)
3–5	1,243 (17.3)	4,947 (16.4)	2,138 (17.3)	723 (13.5)	10,576 (16.3)	119,728 (20.3)
5–10	2,380 (33.2)	10,193 (33.8)	2,798 (22.6)	1,749 (32.8)	19,823 (30.6)	115,015 (19.5)
10–15	976 (13.6)	4,375 (14.5)	983 (8.0)	985 (18.4)	8,372 (12.9)	24,311 (4.1)
>15	715 (10.0)	3,412 (11.3)	806 (6.5)	973 (18.2)	6,945 (10.7)	10,758 (1.8)
Total	7,177 (100.0)	30,168 (100.0)	12,355 (100.0)	5,340 (100.0)	64,854 (100.0)	589,728 (100.0)

a. Figures in parentheses are percentages.
Source: Cabinet Statistics Bureau, *Rōdō tōkei jitchi chōsa hōkoku* [Report of a Survey on Labour Statistics] (1924).

(3) Educational Background

Coal-miners, both men and women, had low educational backgrounds (table 2.7). The lack of formal education of women is particularly conspicuous. About 70 per cent of women miners had either not gone to school at all or had not finished primary school. The contrast with the educational level of textile workers is striking. Taking into consideration the age range of the women in the two industries, this discrepancy is to be expected. (The younger the women, the more educated they were.) From these figures, we can see that the social status of women miners was very low. The stigma that marked coal-miners as criminal elements (*gezainin*) and social dropouts throughout the history of modern Japan is probably related to their low education level.

(4) Areas of Recruitment

From where and from what levels of society were these low-class miners recruited? Although it is difficult to distinguish the men from the women in the materials available, the information has been tabulated for the miners of Chikuhō, where the family recruitment pattern was dominant (table 2.8). The largest number of Chikuhō miners was recruited from Fukuoka Prefecture, where the mines were located, then from the neighbouring prefectures in Kyushu, and finally from Shikoku and south-west Honshu. From the information available on the counties and prefectures from which large numbers of miners were recruited, we are able to tell whether the rate of tenancy was high, how fair tenant–landowner relations were, and what the productivity and rice yield of the area was. The proportion of miners who were tenants and impoverished farmers from poor farming areas was high. Whole families had uprooted themselves and gone to work in the mines. What was it that attracted them to this harsh work?

(5) Wages and Expenses

Women haulers and sub-haulers earned 1.5 to 2 times what women in textiles were making. The figures for 1906 show that the average daily wage of women in silk mills and cotton-spinning factories was 23 sen, and in 1924 the former earned 96 sen, and the latter 1 yen 10 sen (table 2.9). But surface workers earned less than the textile workers (see the 1906 and 1924 figures). Thus there was a difference between the wages of the underground and surface workers. A graphic representation of the wage differentials is shown in table 2.10.

If the shorter working hours of women miners resulting from the demands made on them in the home are taken into account, the difference in the monthly and annual incomes of underground workers and surface and textile workers is much less. But of the women who worked in the mines, those who performed the job of pitman received the highest income. It was for the high pay that the women went into the mines, despite the hardship. Their educational backgrounds marked them as low-class citizens, but they earned the highest wage among women workers.

Table 2.7. Education of Miners, 1924

Education[a]	Female textile workers	Male metal/machine workers	Male miners	Female miners
No schooling	38,903 (6.5)[b]	6,977 (2.7)	28,299 (15.7)	23,115 (35.5)
Elementary, partial attendance	105,216 (17.6)	23,338 (9.0)	46,655 (25.9)	22,534 (34.6)
Primary, graduates	409,994 (68.6)	97,818 (37.9)	62,087 (34.5)	16,718 (25.7)
Upper elementary, partial attendance	14,351 (2.4)	20,614 (8.0)	10,378 (5.8)	1,028 (1.6)
Upper elementary, graduates	16,138 (2.7)	81,607 (31.6)	28,610 (15.9)	1,481 (2.3)
Vocational, partial attendance	2,572 (0.4)	3,936 (1.5)	244 (0.1)	27 (0.0)
Vocational, graduates	1,123 (0.2)	7,118 (2.8)	228 (0.1)	24 (0.0)
Middle, partial attendance	3,320 (0.6)	8,781 (3.4)	1,561 (0.9)	40 (0.1)
Middle, graduates	523 (0.1)	5,630 (2.2)	495 (0.3)	25 (0.0)
Other	5,393 (1.0)	2,590 (1.0)	1,429 (0.8)	155 (0.2)
Total	597,533 (100.0)	258,409 (100.0)	179,986 (100.0)	65,147 (100.0)

a. Elementary school represents grades 1–4, and upper elementary school grades 5–6; middle school is equivalent to high school under the post-war education system.

b. Figures in parentheses are percentages.

Source: Cabinet Statistics Bureau, *Rōdō tōkei jitchi chōsa hōkoku* [Report of a Survey on Labour Statistics] (1924).

Table 2.8. Home Prefectures of Chikuhō Miners, July 1928

Prefecture[a]	County[b]	Miners	Tenancy rate (%)	Rice yield (in *koku*)
Fukuoka (28,280 miners, 44.4%)	Tagawa	6,713	58.3	1.95
	Kaho	5,650	52.1	1.87
	Kurate	3,981	55.3	1.75
	Asakura	1,768	47.2	2.00
	Onga	1,447	50.4	1.79
Prefectural average		—	49.7	2.04
Kumamoto (5,973 miners, 9.4%)	Aso	823	27.0	1.69
	Yatsushiro	743	54.3	1.82
	Tamana	676	52.5	2.02
	Amakusa	520	43.6	1.17
	Ashikita	516	46.8	1.17
Prefectural average		—	43.7	1.79
Oita (5,933 miners, 9.3%)	Hita	1,149	51.4	1.85
	Usa	767	51.8	1.94
	Shimoge	691	38.6	1.90
	Hayami	597	40.3	1.56
	Kusu	526	50.6	1.74
Prefectural average		—	41.1	1.80
Hiroshima (4,278 miners, 6.7%)	Takata	646	42.3	1.74
	Kamo	458	38.4	1.57
	Yamagata	457	35.2	1.55
	Hiba	399	42.9	1.96
	Futami	371	40.8	1.88
Prefectural average		—	39.8	1.76
Ehime (3,337 miners, 5.2%)	Nii	625	72.0	1.87
	Kami Ukena	507	34.9	1.67
	Shūsō	404	53.6	2.15
	Uma	338	47.6	2.14
Prefectural average		—	42.5	2.01
Saga (2,853 miners, 4.5%)	Saga	533	45.8	2.44
	Miyaki	528	50.6	2.14
	Kanzaki	423	37.9	2.16
	Ogi	327	46.4	2.14
Prefectural average		—	41.1	2.09

Table 2.8. (*continued*)

Prefecture[a]	County[b]	Miners	Tenancy rate (%)	Rice yield (in *koku*)
Kagoshima (2,202 miners, 3.5%)	Aira	447	52.0	1.66
Prefectural average		—	38.5	1.52
Nagasaki (1,494 miners, 2.3%)	Nishisonoki	302	29.0	1.43
Prefectural average		—	35.2	1.67
Shimane (1,317 miners, 2.1%)	Ochi	405	?	1.76
	Naka	314	?	1.54
Prefectural average		—	?	1.89

a. Figures in parentheses represent number and percentage of miners at Chikuhō from that prefecture. Not included are: Koreans (5,626, 8.8 percent), Miyazaki Prefecture (1,305 miners, 2.1 per cent), and Yamaguchi Prefecture (1,048 miners, 1.6 per cent).

b. For Fukuoka, Kumamoto, and Ōita, the five counties from which the largest number of miners came are tabulated; for other prefectures, only counties with over 300 miners are included.

Source: Compiled from Fukuoka Chihō Shokugyō Shōkai Jimukyoku, *Chikuhō tankō rōdōsha shusshinchi shirabe* [Survey of Home Prefectures of Chikuhō Miners] 1931; *Kakuken tōkeisho* [Statistics by Prefecture] (1928). For Shimane, *Kakuken tōkeisho* (1927) was used.

Table 2.9. Average Daily Wages of Women Miners, 1906 and 1924 (yen)

	1906				1924 (June)			
					Underground		Surface	
Mine Average	Miner	Sub-hauler	Bracer	Dresser	Digger	Other	Worker	Average
Jōban	0.350	0.304	—	0.189	1.845	1.180	0.666	1.359
Chikuhō	0.551	0.492	0.508	0.278	1.834	1.269	0.739	1.390
Karatsu	0.466	0.382	0.430	0.216	1.817[a]	1.338[a]	0.664[a]	1.383[a]
Miike	—	0.404	0.240	0.202				

a. All Kyushu excluding Chikuhō.

Source: Compiled from Agriculture and Commerce Ministry, Bureau of Mining, *Kōfu taigū jirei* [Examples of Treatment of Miners] (1908), p. 57; Nihon Ginkō Chōsa Kyoku, *Kōzan rōdō tōkei* [Statistics on Labour in Miners] (June 1924).

Table 2.10. Household finances of a Mining Family, 1925 (yen)

		Hewers (incl. fixers)[a]			Transporters (incl. pole carriers)		Factory workers	
		Only household head works	One woman from household works: Hauler	One woman from household works: Other	Only household head works	One woman from household works	Only household head works	One woman from household works
Income	Wages							
	Household head	63.23	56.19	54.55	49.66	47.04	61.83	53.87
	Family	—	32.77	27.98	—	18.51	—	18.29
	Subsidies/relief aid	1.82	2.84	2.79	0.69	0.94	0.05	0.11
	Bank balance	13.76	16.60	15.15	8.49	11.00	15.61	9.71
	Bank withdrawals/loan collection	5.23	6.92	5.81	3.64	5.80	6.77	5.55
	Loans/pawns	3.02	3.73	2.93	2.92	2.01	1.54	1.33
	Total (incl. misc.) (A)	89.87	120.46	111.36	69.31	86.49	90.14	90.85
Expenses	Work equipment	1.80	2.95	2.39	0.54	0.84	0.48	0.84
	Food	31.12	36.13	35.78	28.68	30.83	30.21	33.05
	Housing	3.79	4.11	3.94	2.50	3.06	2.60	3.68
	Clothing	6.12	7.08	6.98	3.92	6.00	6.52	7.75
	Education/medical expenses	3.18	6.59	5.75	2.57	4.46	3.62	5.47
	Entertainment	9.51	10.73	9.99	7.06	7.71	9.11	8.99

Savings	9.12	15.40	13.86	6.21	8.50	9.44	9.11
Loan payments	3.46	6.36	4.73	5.64	6.12	5.12	6.96
Total (incl. misc.) (B)	73.70	96.36	89.88	60.91	73.05	73.29	80.98
Difference (A–B)	16.17	24.10	21.48	8.40	13.44	16.85	9.87
Engel coefficient	42.2	37.5	39,8	47.1	42.2	41.2	40.8
No. of households surveyed	80	70	33	47	42	73	41

a. Fixers are miners who fix tunnels and hewing sites.

Source: Compiled from Shakai Kyoku, Sekitan Kōgyō Rōdō Jijō Chōsakai [Social Affairs Bureau, Study Group on Labour Conditions in the Coal-mining Industry], *Tankō kōfu kakei chōsa* [Survey of Household Finances of Miners] (1925).

The household expense budget was greatly influenced by the income of a woman working underground. The household budget of relatively high-income miner families in large mines has been tabulated (table 2.10). None of the families can cover their expenses with their monthly incomes; savings, loans, and pawns provide a sizeable proportion of income. The high annual savings rate seen under expenses does not indicate a surplus in funds. These were mostly forced savings, enforced by the mine operators to stop miners from leaving.[11] All the households were on a tight budget, but those with women haulers had the highest incomes. These women helped boost their family incomes by 27.2 per cent. Their Engel coefficient and clothing, social and entertainment expenses indicate that their families enjoyed a more comfortable life than other mining families.

In sum, women miners can be characterized as middle-aged, long-term, skilled workers when they worked underground as haulers and supporters; as surface workers, they were young and did not stay in their jobs for long. Notwithstanding their social status, underground workers endured debilitating hardships for a standard of living that was, in relative terms, higher than that of their counterparts in the textile industry.

II. Technological Innovations and Women Workers

1. Background Factors

The advances made during the industrial revolution in the mechanization of the coal industry were only in the conveyance system. They had stopped short of revolutionary changes in the hand tools used to hew coal. Full-scale mechanization of the actual mining process did not begin until the 1920s, particularly from the latter half of the decade.

First, let us examine the historical factors that helped usher in the innovations. The chronic recession of the 1920s caused a drop in the market price of coal (table 2.11). For example, in 1920, Type 1 Kyushu coal (Moji) cost 28.55 yen per ton, and a year later had dropped to 20.20 yen. Following the establishment of the Federation of Coal Industries in 1921, restrictions were placed on the amount of coal sent, and the price per ton settled at over 16.00 yen.[12] But the Tokyo prices for Type 1 Kyushu coal, Type 1 Iwaki coal, and Yūbari coal continued to fall in 1922 and after, placing the whole industry in severe financial straits.[13] The situation deteriorated with competition from imported coal. As the figures in table 2.11 show, coal imports rose from 1922, and in 1923 and 1924, and from 1927 onward they exceeded coal exports. Mining was an industry that had grown to meet domestic demands and had tried to meet export requirements as well. But it reached a turning-point when coal began to be imported in large volume. The imported coal, which came mainly from Guandong Province (particularly the Fushun mine), had been mined by cheap colonial labour and the easier mining conditions of surface mining.[14] The competition brought by low-cost

Table 2.11. Domestic Coal Yield and Quantity of Exported/Imported Coal, 1914–1929 (1,000 tons)

Year	Yield	Export	Import	Import sources			
				Rep. of China	Kwangtung Province	French Indochina	Other
1914	22,293 (100)[a]	3,589 (100)	958 (100)	701	191	48	17
1915	20,491 (92)	2,923 (81)	615 (64)	426	67	107	14
1916	22,902 (103)	3,017 (84)	556 (58)	315	111	124	7
1917	26,361 (118)	2,813 (78)	713 (74)	490	125	94	4
1918	28,029 (126)	2,197 (61)	768 (80)	539	136	88	5
1919	31,271 (140)	2,017 (56)	705 (74)	467	125	108	5
1920	29,245 (131)	2,147 (60)	810 (85)	567	87	152	4
1921	26,221 (118)	2,407 (67)	790 (82)	418	214	157	2
1922	27,702 (124)	1,704 (47)	1,187 (124)	410	575	184	18
1923	28,949 (130)	1,587 (44)	1,713 (179)	647	735	196	134
1924	30,111 (135)	1,725 (48)	2,012 (210)	577	1,139	192	104
1925	31,459 (141)	2,698 (75)	1,768 (185)	302	1,285	168	13
1926	31,427 (141)	2,611 (73)	2,045 (213)	311	1,418	244	71
1927	33,531 (150)	2,191 (61)	2,703 (282)	560	1,743	346	54
1928	33,860 (152)	2,185 (61)	2,779 (290)	537	1,759	377	107
1929	34,258 (154)	2,044 (57)	3,254 (340)	628	2,016	488	123

a. Figures in parentheses based on 1914 as 100.

Source: Compiled from *Nihon kōgyō hattatsushi* [History of the Development of Coal-mining in Japan], vol. 2 (Kōzan Konwakai, 1932), pp. 174–75, 181–82.

imported coal pushed the domestic industry into a quagmire. As the recession worsened, the competition intensified. This was the first major factor that helped accelerate the technological renovation of the industry.

The second factor was protective legislation for coal-miners. As a participant in the first International Labour Conference (October–November 1919), Japan undertook to adopt the agreements made by reforming its 1916 legislation on miners, the Regulations concerning Relief for Coal-miners. This was one of the most important labour issues that the government acted on in the post-First World War era. The important reforms were the prohibition of late-night labour and underground work by women and minors. These reforms undermined the basic structure of the workforce—the working couple and family recruitment. For the mine operators, for whom this mean further financial restraints, the reforms were unacceptable. The union of coal mines in Chikuhō started an opposition movement against the reforms.[15] But from the viewpoint of owners of large mines, the removal of women and children from the mines became an opportunity to implement technological changes to raise productivity and cut costs.

On 1 September 1928, the reform of the Regulations concerning Relief for Coal-miners was adopted by the Home Ministry as Order No. 30. Although there was a five-year period of grace before its final enforcement, the reforms helped accelerate the implementation of technological reforms. The mines that had been dependent solely on family and couple work teams were forced to revamp their employment patterns, raise their productivity, and improve their technology.

There were two important exceptions to the 1928 reform. One concerned the prohibition of night work. The surface workers, the coal dressers, could work at night for the time being provided there were three shifts a day, and that this was for a predetermined period of time. For the mining of thin seams, women and children were allowed to work underground with the permission of the mine superintendent.[16]

Secondly, as the enforcement of the reforms approached the final year of grace, there was a movement to postpone the date by the smaller and medium-sized mines (Coal Mine Mutual Aid Association). The Home Ministry consequently issued another order (5 June 1933), which allowed underground work by women and children, mainly in pits with limited reserves.[17]

2. Technological Reforms

The first major reform involved the mining method—from the pillar method to the longwall method (fig. 2.4). The pillar method required boring directly into the seams, leaving pillars in the resulting holes as support. The longwall method involved mining the coal along the whole seam at the angle it was in, while leaving pillars for support. Eventually, the pillars were eliminated, and the *sobarai* longwall method was employed. This had been used earlier during the industrial revolution, in mines with thin seams where

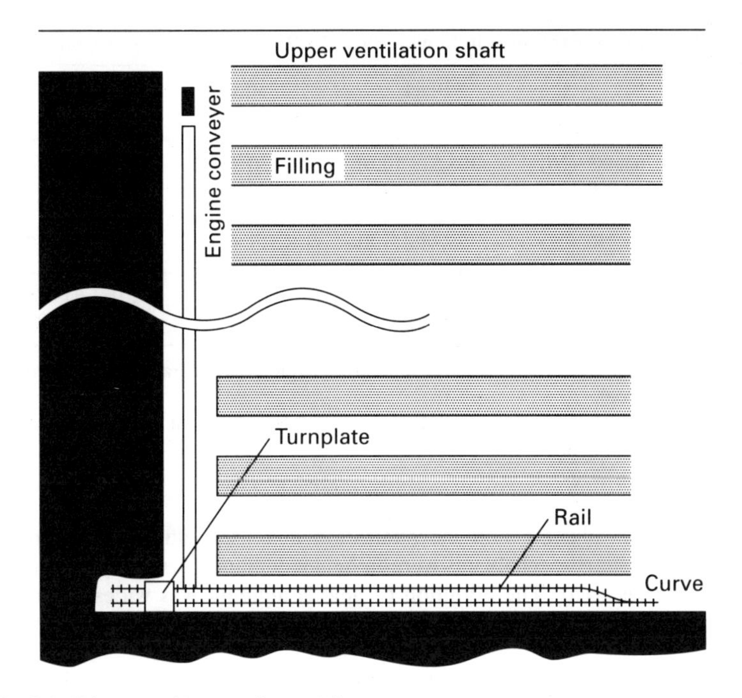

Fig. 2.4. Diagram of Longwall-type Mine
Source: Compiled from *Nihon kōgyō hattatsushi* [History of the Development of Coal-mining in Japan], vol. 2 (Kōzan Konwakai, 1932).

geological pressures were controllable. Improvements in filling made it possible to use longwall mining in thick seams as well as piled seams. This changed the nature of the work from small teams working in isolation to one in which greatly enlarged working space, group work, and the use of machines became possible.[18]

Another improvement in mining techniques was in blasting. Explosives had been used primarily to wedge thin spoil and petrified roots from the seams.[19] The widespread use of safer explosives (table 2.12) following the Second World War and the adoption of boring machines and drills firmly established blasting as a technique. Before the boring machine, it would take 20 to 30 minutes to bore one hole 90 cm deep, and now it took just a few minutes to bore a hole 120 to 150 cm deep; where one blast had opened 20 to 40 holes, now 20 to 40 could be bored.[20]

Thirdly, the coal cutter and pick were made obsolete by the adoption of mechanized tools. The adoption of machinery was extremely difficult when the pillar method was used and working space was severely limited; but with the longwall method, which was becoming increasingly popular, mechanization became possible.

Table 2.12. Use of Explosives in Mining, 1923–1929

Year	Dynamite	Black powder	Cotton powder	Safety explosives
1923	763,483 (26)[a]	105,518 (3.6)	11,430 (0.4)	122,636 (4)
1924	769,788 (26)	79,855 (2.7)	10,798 (0.4)	497,641 (17)
1925	812,856 (26)	53,598 (1.7)	13,688 (0.4)	745,913 (24)
1926	693,230 (22)	55,386 (1.8)	8,513 (0.3)	977,914 (31)
1927	708,927 (21)	33,605 (1.0)	2,188 (0.1)	1,453,136 (43)
1928	654,731 (19)	23,979 (0.7)	1,507	1,954,843 (58)
1929	669,503 (20)	20,668	?	3,822,314 (112)

a. Figures in parentheses indicate the amount of explosives (in grams) used to extract one ton of coal.

Source: *Nihon kōgyō hattatsushi* [History of the Development of Coal-mining in Japan], vol. 2, table 60.

Fourthly, as blasting and mechanization increased efficiency, improvements in transport mechanisms were in heavy demand. A variety of hauling apparatus was tried. The adoption of new equipment was widespread and immediate, as it was more efficient and cheaper by far than the woman haulers.[21]

Other innovations included the switch from steam to electric power in slope mining, and use of a coal-washing machine to enhance the quality of coal production for the market.

The degree to which technological innovations were adopted varied from mine to mine. The number of new mining machines acquired nationwide rose at dramatic rates in 1926, 1927, and 1933 (see table 2.13). The peak years for the acquisition of new machinery were, by prefecture, 1926 for Hokkaido and Fukuoka, 1928 for Nagasaki, and 1931 to 1933 for Fukushima, Ibaraki, and Yamaguchi. The discrepancy between prefectures with the most new machinery (Hokkaido) and those with the least (Fukushima, Ibaraki, and Yamaguchi) was great, an indication of the regional differences between the technologically advanced Hokkaido and the three other prefectures, which were slow developers. This was also seen in the productivity per miner (fig. 2.5): from 1920 on, there was a dramatic in rise productivity in Hokkaido, Nagasaki, and Fukuoka; for Fukushima, Ibaraki, Yamaguchi, and Saga, there was a slow rise; and for Ibaraki, there was actually a drop in productivity from 1933 on.

The size of the coal mines was one of the reasons for this discrepancy. Hokkaido mines were generally large, while those in Fukushima, Ibaraki, and Yamaguchi were proportionally small- and medium-sized. One can see that the level of mechanization was directly linked to the economics of each mine and the extent of capital outlay. This is apparent in the case of Chikuhō: new machinery was concentrated in the big-capital mines run by Mitsui, Mitsubishi, Meiji, and Kaijima Gomei. This is reflected in the productivity

Table 2.13. Installation of Coal-mining Machinery (rock drills, augers, coal cutters, coal picks) by Prefecture, 1925–1935

Prefecture	1925	1926	1927	1928	1929	1930	1931	1932	1933	1934	1935
Machines installed											
Hokkaido	44	102	111	145	198	300	228	238	500	659	896
Fukushima, Ibaraki	30	29	39	6	24	16	70	44	88	107	129
Yamaguchi	1	—	—	5	2	16	—	4	47	27	41
Fukuoka	32	152	176	461	429	341	190	378	520	667	619
Sago	57	8	138	46	45	21	—	58	34	55	40
Nagasaki	6	—	5	66	100	82	160	96	138	190	119
Nationwide total	170	291	467	730	861	776	648	818	1,327	1,749*	1,850*
Machines installed per mine[a]											
Hokkaido	1.47	3.64	3.83	5.18	7.07	10.34	6.91	6.61	14.29	17.34	24.22
Fukushima, Ibaraki	1.20	1.45	1.86	0.29	1.14	0.84	3.33	2.00	4.00	4.65	5.16
Yamaguchi	0.09	0	0	0.45	0.14	1.23	0	0.36	3.92	1.93	2.93
Fukuoka	0.41	1.92	2.23	5.84	6.74	4.87	2.50	4.91	6.75	8.23	7.11
Sago	5.70	1.00	17.00	5.88	6.43	3.50	0	9.67	5.67	7.86	6.67
Nagasaki	0.35	0	0.29	3.14	4.76	4.32	8.89	5.33	6.90	7.60	4.41
Nationwide total	0.96	1.80	2.76	4.32	5.22	4.91	3.90	4.81	7.72	9.25	9.34

a. Figures for machines per mine calculated by dividing the number of machines installed by the number of mines that yield more than 10,000 tons annually.

Source: Compiled from Commerce and Industry Ministry, Mining Bureau, *Honpō kōgyō no sūsei* [Mining Trends in Japan].

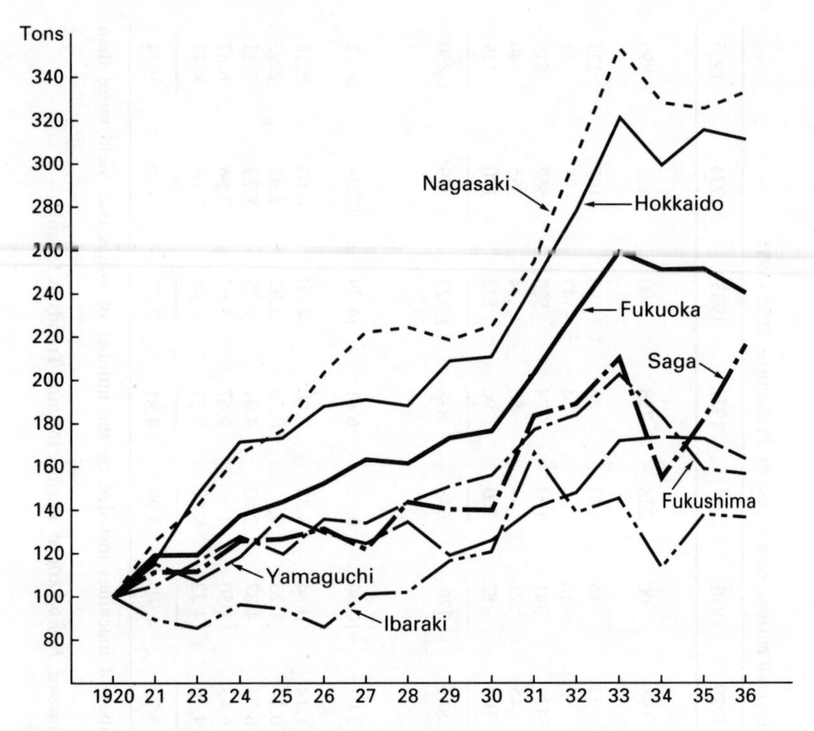

Fig. 2.5. Trends in Mine Productivity per Miner, 1920–1936 (1920 = 100)
Note: Figures for 1922 are unknown.
Source: Compiled from Commerce and Industry Ministry, Mining Bureau, *Honpō kokyō no sūsei* [Mining Trends in Japan].

of the mines (fig. 2.6); from the second half of the 1920s and into the 1930s, the large mines established supremacy in productivity over the smaller mines.[22]

The machines that were purchased by the mines, particularly from the latter 1920s, were mostly imported. In 1932, only 20 per cent of the machines were domestic products, and the mines thus depended on foreign machines for the most part. The technological revolution apparently had little impact on the domestic machine industry. But from 1933, a sudden increase in machine production is seen, and by 1935 the country was 50 per cent self-sufficient.[23] There were three major types of machinery manufacturers: the large manufacturers, such as Hitachi Seisakujo and Komatsu Seisakujo; the companies that ran mines as well as manufacturing companies, such as Sumitomo Kikai Seisakujo, Miike Seisakujo, and Ashio Seisakujo; and the small- and medium-sized manufacturers (including companies that specialized in mining machinery) (table 2.14). From 1933 on, a

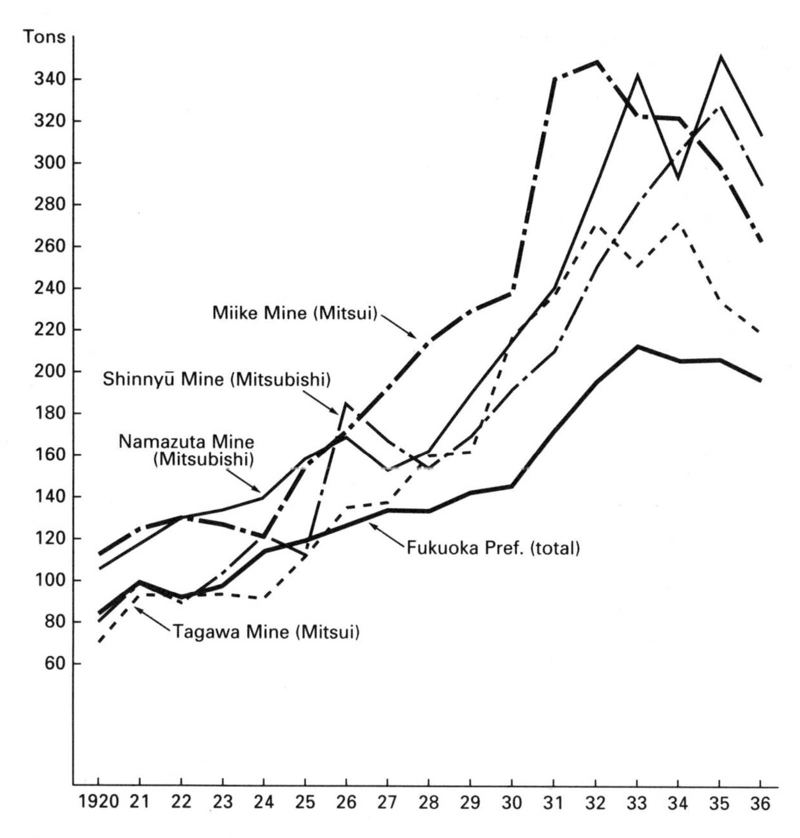

Fig. 2.6 Trends in Productivity per Worker in Fukuoka Prefecture, 1920–1936
Note: Calculated by dividing output by the number of miners. Figures for Fukuoka for 1922 are unknown.
Source: Compiled from Commerce and Industry Ministry, Mining Bureau, *Honpō kōgyo no sūsei* [Mining Trends in Japan] for each year; *Mitsubishi kōgyō shashi* [History of the Mitsubishi Mining Company] (1976), documents, tables 14 and 15; and *Mitsui kōzan gojūnen shikō* [Manuscript for 50-year History of the Mitsui Mines], vols. 5–2 and 16.

complementary relationship developed as technological advances encouraged domestic machine manufacturers, and vice versa.

3. The Decline of Women Workers

The prohibition of late night and underground work by women and minors by the reform of the Rules for Relief of Miners and the rationalization of the work process brought about by technology drastically influenced the reduction of women workers. As the statistics in table 2.15 indicate, the pro-

Table 2.14. Domestic Production of Mining Machinery, 1936

Company	Workers as of Oct. 1936[a]	Rock drill, auger	Coal-cutter	Pneumatic hammer, pick
Hitachi Seisakujo	7,261	—	50	—
Komatsu Seisakujo	1,095	195	—	—
Sumitomo Kikai Seisakujo	1,163	4	—	—
Miike Seisakujo		325	41	—
Ashio Seisakujo		206	—	144
San'ei Seiki Seisakujo	181	234	—	177
Taisei Seisakujo	117	254	—	136
Kaneshiro Sakuganki Seizō	149	297	—	—
Kurita Seisakujo		112	—	42
Yamamoto Tekkōjo	105	135	—	200
Uryū Seisakujo		57	—	103
Nihon Kūki Kikai Seisakujo		60	—	154
Seikōsha	62	49	—	—
Teikoku Sakuganki Seisakujo		48	—	7
F.K. Seisakujo	76	73	—	—
Kyōritsu Kikai Seisakujo	79	—	—	28
Nihon Sakuganki Seisakujo	68	5	—	—
Nakayama Kōgyōshō	114	7	—	—
Seibu Denki Kōgyōsho	164	475	—	—
Ōsaka Tokki Seisakujo		142	—	—
Chiyoda Seisakujo	75	28	—	—
Other (4 manufacturers)		45	—	—
Total		2,751	91	1,525

a. Columns are blank when information unavailable.

Source: Compiled from Commerce and Industry Ministry, Mining Bureau, *Honpō kōgyō no sūsei* [Mining Trends in Japan] (1936); figures for workers are from Kyōchōkai Sangyō Fukuribu, *Zenkoku kōjō kōzan meibo* [Directory of Factories and Mines in Japan] (1937).

portion of women starts to decline in 1925, and 1928 and 1931 were years in which the decrease was considerable. The drop was especially prominent in underground workers: in the 11 years between 1920 and 1931, the proportion fell by 19 per cent.

The labour system entirely dependent on a family recruitment system was fast disintegrating. But even after 1934, the year in which the relief measures prohibiting underground work went into effect, 4 per cent of the pitmen were still women—which was, as discussed above, possible because of the exceptions that were allowed. The decrease in women working on the surface was not as noticeable, and in the 17 years between 1920 and 1936 the percentages dropped by only 5 points. This is partly because the technological innovations for surface work did not progress so rapidly and part-

Table 2.15. Women Miners, 1920–1936

	Women miners			Percentage of women miners[b]		
Year[a]	Underground	Surface	Total	Underground	Surface	Percentage of workforce
1920	66,396	28,474	94,870	26.6	30.6	27.7
1921	50,695	21,927	72,622	26.3	29.2	27.1
1923	54,898	22,151	77,049	26.8	30.0	27.6
1924	47,948	20,504	68,452	26.2	30.0	27.3
1925	46,072	19,330	65,402	24.8	28.8	25.9
1926	42,214	18,815	61,029	24.6	29.8	26.0
1927	41,701	18,039	59,740	23.6	29.0	25.0
1928	37,730	16,947	54,677	21.3	27.9	23.0
1929	32,977	16,300	49,277	19.4	27.8	21.5
1930	24,002	13,438	38,440	15.8	27.4	18.8
1931	10,992	10,599	21,591	9.8	24.9	14.0
1932	7,202	9,420	16,622	7.2	24.6	12.0
1933	6,573	9,437	16,010	6.3	24.5	11.1
1934	6,598	12,327	17,925	4.5	27.3	10.6
1935	5,308	12,539	17,847	4.1	26.8	10.2
1936	4,841	13,427	18,268	3.3	25.9	9.2

a. Figures for 1922 not available.
b. Percentages calculated by dividing number of women workers by total workers.
Source: Compiled from Commerce and Industry Ministry, Mining Bureau, *Honpō kōgyō no sūsei* [Mining Trends in Japan].

ly because late-night work, as long as a three-shift work schedule was used, was permitted.

There were regional differences in the decline in women miners (fig. 2.7), closely following the pattern of regional differences in implementation of new technology. In the mines where the slope transport system was prevalent, women haulers were almost non-existent. In Hokkaido, the proportion of women workers, which was low from the outset, began gradually to decrease further in the 1920s. Fukuoka and Saga prefectures witnessed dramatic drops in 1928 and 1929, while in Nagasaki, Fukushima, Ibaraki, and Yamaguchi, the decrease was not as great. In 1931, Ibaraki Prefecture hit its all-time low, and from 1932 increased again. We can conclude that in areas where mine operations were run on a small scale (Fukushima, Ibaraki, Yamaguchi), the implementation of labour reforms, i.e. the curtailment of female labour, had not been thorough.

This can been seen in by comparing the percentages for the decline of women miners in large mines and in small and medium mines in the Chikuhō area (table 2.16). Because specialization was slow in developing in the smaller mines, the proportion of workers underground averaged a high 60

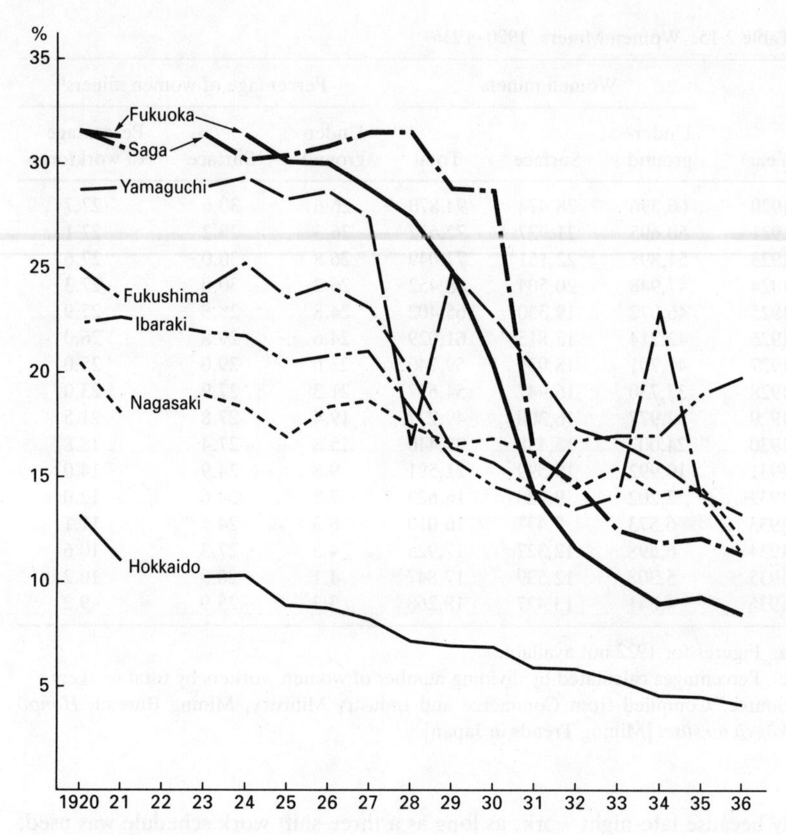

Fig. 2.7. Fluctuations in Proportion of Women Miners in Hokkaido and Other Prefectures, 1920–1936
Note: Figures for 1922 unknown.
Source: Compiled from Commerce and Industry Ministry, Mining Bureau, *Honpō kōgyō no sūsei* [Mining Trends in Japan].

per cent. In 1930, this drops, but in the following year, the proportion increases to the 60 per cent range again. In contrast, sharp declines are seen in the following mines: from 1924 in Hōkoku, from 1928 in Tagawa and Ōnoura, and from 1930 in Namazuta and Tadakuma. Because of their inability to join the technological revolution the smaller mines were still dependent on women, and had not eliminated female labour at the rate the large mines had.

The legal restrictions and the technological revolution that contributed to the decline of women miners have been examined, but the following two developments need to be mentioned as well. The first is that during the industrial revolution, when recruitment of women workers and whole families

Table 2.16. Hewers and Haulers, Chikuhō Area, 1920–1932[a]

Year	Tagawa (Mitsui)	Namazuta (Mitsubishi)	Tadakuma (Sumitomo)	Ōnoura (Kaijima)	Toyokuni (Meiji)	Small mines with 100–499 workers (no. of mines)[b]
1920	36.4	62.4	54.2	37.9	44.0	55.3 (13)
1921	38.5	61.5	51.2	34.9	48.7	63.8 (14)
1922	38.9	65.3	53.7	43.7	48.5	66.4 (7)
1923	39.1	61.6	57.1	42.1	47.3	58.1 (5)
1924	38.1	60.4	56.6	36.9	39.9	60.0 (17)
1925	39.6	59.5	58.2	36.3	40.6	62.5 (14)
1926	40.3	53.4	57.7	33.9	41.9	63.7 (14)
1927	39.7	59.0	56.3	36.7	41.9	65.9 (9)
1928	33.3	61.6	55.3	32.0	39.1	66.6 (22)
1929	31.0	57.4	56.9	32.4	36.6	68.5 (20)
1930	28.0	41.1	47.0	31.4	33.4	59.5 (9)
1931	23.9	42.8	45.6	26.6	32.4	63.2 (14)
1932	22.8	37.6	46.4	24.4	34.7	60.0 (21)

a. Figures for 1919–1923 are as of end of May; for 1929, end of June; all others, except Tagawa, as of end of September.
b. Figures in parentheses are percentages.
Source: For Tagawa, compiled from *Mitsui kōzan gojūnen-shi kō* [A Fifty-year History of the Mitsui Mines], vol. 16, "Labour," Mitsui Bunko; for the rest, *Chikuhō sekitan kōgyō kumiai geppō* [Monthly Newsletter of Chikuhō Miners' Union].

was prevalent, the *naya* system came into being to accommodate the situation, but with the demise of the female labour force, this system declined, and the management of labour was undertaken directly by the mine operators. In 1922, for example, it was decided at a Mitsubishi Kogyo meeting of work supervisors that the *sewanin* system (a subcontractor-supervisor system similar to the *naya* system) would be abolished. In February 1929, in the mines at Shinnyū, Namazuta, Hōjō, and Kamiyamada, which were all managed by Chikujo Kōgyōsho, 43 *sewanin* and 39 wakers (*hitoguri*), whose job it was to wake the miners up and encourage them to go into the mines, lost their jobs. In March, Ōchi and Yoshinotani mines, run by Karatsu Kōgyōsho, and, in August, Iizuka Kōgyōsho followed suit.[24] And in October 1929, even Sumitomo's Tadakuma mine, which had prided itself on having the oldest and most authentic *naya* system in the Chikuhō area, fired 14 stable foremen, 27 subforemen, 17 wakers, and 13 accountants working for the foremen.[25] Other mines in Kyushu (such as the Shakano'o and Shimoyamada mines of Furukawa Kōgyō and the Yoshikuma Tsunawake, Mameta, and Yoshio mines of Aso Shōten) dismissed their foremen in 1929, and the *naya* system almost completely disappeared from the major mines of Kyushu.[26]

There are two factors that brought about the end of the *naya* system. The first concerns work organization. Under the longwall method, the task of hewing coal became a cooperative job, and the mechanization of the process changed the nature of the work. When the pillar method was used, work units were isolated and dispersed; their work was done entirely by hand, and supervision of the workers by foremen who patrolled the mines was necessary. New technology spelled the end for women workers, but at the same time it also forced a change in labour management. Secondly, as women were being excluded from their jobs, mechanization decreased the demand for a large workforce. The emphasis shifted to enhancing the quality of the workforce. The recruiting methods of the *naya* foremen, who had hired indiscriminately, became obsolete.

The second development is the increase in Korean miners that accompanied the elimination of women miners. There was a high concentration of Korean miners in Fukuoka Prefecture, as can be seen from table 2.17.[27] Eighty-six per cent of them had underground jobs, the important part of mining work. What made this possible was the longwall method and the change to team work involving simultaneous work by many miners, as opposed to the old units of two or three working separately. A survey from Mitsubishi Namazuta mine reports: "Koreans are patient, strong workers. Although they are untrained, they can be hired for team work." "The adoption of the longwall system eliminated jobs for haulers, and thus couples could not work together. Where two incomes could not be had, with the woman's job eliminated, Japanese men seeking mine work declined."[28]

Thus the adoption of the longwall system contributed to the exclusion of women haulers and brought in team work; and this in turn made the recruitment of Koreans possible. One other fact can be gleaned from table 2.17: the Koreans were unevenly distributed in the mines. They were numerous in the four Mitsubishi Kōgyō mines of Shinnyū, Namazuta, Kamiyamada, and Hōjō and at Iizuka, which was also a Mitsubishi-affiliated mine. Eighty per cent of Fukuoka Korean miners worked in these five mines, and represented a large proportion of the workforce there: over 30 per cent in Shinnyū, Namazuta, and Iizuka. In a sense, the Korean miners were hired at a time of technological change, and they did replace the women, but this is not the only factor. Circumstances surrounding each mine, e.g. the labour practices and the situation of coal dispursement (the degree of difficulty in extracting the coal—also played a role in their recruitment.

We have discussed the decline of women workers and the relationship this had with the decline of the *naya* system and the hiring of Koreans. The question of why there was no resistance to the elimination of women from the workforce still remains unanswered. There was not a single labour dispute involving the firing of women. The women underground workers earned a substantial amount of the mining family's income, and it is difficult to understand why there was no resistance.

Two pertinent factors should be examined. First, in order to assist the miners whose incomes were suddenly to fall, the women who had been dis-

Table 2.17. Korean Miners by Mine in Fukuoka, March 1928

Mine	All miners			Korean miners			Percentage of Korean miners		
	Underground	Surface	Total	Underground	Surface	Total	Underground	Surface	% of total workers
Shinnyū (Mitsubishi)	2,102	744	2,846	873	71	944	41.5	9.5	33.2
Namazuta (Mitsubishi)	3,207	1,336	4,543	1,494	244	1,738	46.6	18.3	38.3
Kami-Yamada (Mitsubishi)	2,132	525	2,657	218	121	339	10.2	23.0	12.8
Katashiro (Mitsubishi)	1,926	820	2,746	392	21	413	20.5	2.6	15.0
Iizuka (Nakajima Kōgyō)	4,687	1,114	5,801	1,687	81	1,768	36.0	7.3	30.5
Ōnoura (Kaijima)	6,684	2,828	9,512	58	154	212	0.9	5.4	2.2
Yoshio (Asō)	2,401	594	2,995	48	90	138	2.0	15.2	4.6
Tsunawake (Asō)	761	222	983	110	16	126	14.5	7.2	12.8
Yoshisumi (Asō)	1,273	341	1,614	96	—	96	7.5	—	5.9
Arameo (Pvt.)	112	17	129	115	20	135	102.7	117.6	104.7
Other (21 mines)	—	—	—	513	89	602	—	—	—
Total	—	—	—	5,604	907	6,511	—	—	—

Source: Compiled from Fukuoka Chihō Shokugyō Shōkai Jimukyoku, *Kannai zaijū Chōsenjin rōdō jijō* [Working Conditions of Korean Miners in the District] (1929), and *Chikuō sekitan kōgyō kumiai geppō* [Monthly Newsletter of Chikuhō Miners' Union].

charged were encouraged to work in subsidiary jobs. For example, Mitsui's Miike Kōgyōsho adopted a policy in 1928 to alleviate the pressures on miners by establishing a work centre [29] where a variety of side jobs were encouraged; the products manufactured at the centre were sold on commission through the Miike Cooperative sales department.[30] The types of jobs offered, the numbers of workers, and wages in 1930 are summarized below.[31]

1. Manufacturing of products purchased by the company: average daily wage, 50 sen; 3,518 workers in the past five months.
2. Bamboo mat making: average daily wage, 80 sen; 1,518 workers in the past five months.
3. Sewing-machine work (53 machines available, either borrowed from the company or purchased on instalment): 80 sen per day.
4. Sewing: 35 workers averaging 180 articles of clothing per month; average daily wage 1.13 yen.
5. Weaving: 6 workers; average daily wage 38 sen.
6. Production of dynamite balls: 6 workers; average daily wage, 50 to 60 sen.

Miike Kōgyōsho's work centre expanded in 1932 to become Miike Cooperative Union for Subsidiary Work.[32] The encouragement of subsidiary work was not limited to Miike: other mine companies had similar programmes. In June 1929, the Navy's Shimparu mine, in response to the prohibition of underground work by women, established a Home Industries Committee.[33] In December of the same year, the Onga County Labour Managers' Group, organized by 14 mines in Onga county, Fukuoka, met to deliberate an unemployment policy for women miners. It decided to adopt straw-weaving and pig-farming as subsidiary lines and to launch a campaign for the establishment of a spinning factory.[34] Meiji Kōgyō undertook, in 1930, as an unemployment relief measure for women, a project to expand an athletic field and establish a fish market.[35] In July 1932, in Kaijima Onoura mine, unused housing was remodelled into a factory producing baskets (used for the outer enclosure of mine holes), coal conveyances, fishing nets, and straw mats.[36]

The above examples show that mine operators encouraged multi-occupation families,[37] and that efforts were made to aid the finances of families affected by the dismissal of workers. This progressive policy pursued by the management made for a transition free of conflict as new technology was adopted and women were discharged.

The second reason for the lack of resistance was that the management provided welfare facilities and educational and cultural organizations which were much better than in other industries during the inter-war period. These were established to unify the mine workers. This policy proved to be crucial in checking resistance against the drastic changes that accompanied the implementation of new technology. The mining companies set aside a large proportion of their budget for these activities: the percentage of welfare expenses over total wages was 24.1 per cent in mining (1926) and 23.5

Table 2.18. Chronology of Labour–Management History at Mitsui Kōzan's Miike Mine, 1919–1935

May 1919	Women's group established in miners' housing complex
Sept. 1919	Miike Mines Vocational Night School established
March 1920	Mitsui Miike Mutual Aid Society established
	Youth group established in miners' housing complex
May 1922	Household heads' group established in miners' housing complex
Nov. 1922	Miike Mines Veterans' Association reorganized
Oct. 1924	Boys' and girls' group established in miners' housing complex
Nov. 1924	Workers' school established.
Dec. 1925	Young women's group established at Miike Kōgyōsho
July 1926	Youth Training Centre opens
Jan. 1929	Night school for supplementary education for workers opens
Feb. 1929	Middle-age group established in miners' housing complex
Dec. 1929	Pilgrimage group established; trips to Ise Shrine and Tokyo organized
March 1930	Working Girls' School opens
January 1932	Youth group from housing group reorganized into Mutual Aid Youth Group
June 1933	Miike Kan'yūkai established
Jan. 1935	Branch of Women to Defend the Country League established in housing complex
April 1935	Family Association established for families living outside miners' housing complex

Source: Compiled from *Miike Kōgyōsho enkakushi* [History of Miike Kōgyōsho], vol. 7, Mitsui Bunko.

per cent in spinning—both extremely high figures when other industries did not reach even 10 per cent.[38] All mines surveyed in 1932 had training centres, placing them far beyond other industries.[39] Mining communities were geographically isolated: many of the large mines were closely knit and their activities were self-contained. They built barriers around themselves that stopped the spread of any labour movement into the mines. An example of the way in which policies were implemented for the integration of the workforce at Mitsui's Miike Kōgyōsho is given in chronological form in table 2.18. This chronology demonstrates how the management typically adopted policies embracing the whole community and all family members.

III. Conclusion

The decline in women miners was brought about by the mechanization of the coal-mining process. This was in turn due to a change in labour policy, under the resolutions adopted by the first International Labour Conference. In 1920, the relief measures were implemented as a legal sanction for the protection of coal-miners. The decline in women miners can be seen macroscopically as the beginning of world capitalism, following the First World War, in the framework established by the Versailles Treaty. This suggests that if that framework were eliminated, the phenomenon of women miners could reappear.

In July 1937, after the beginning of the war in China, Japan's economy was placed on a war footing. In the following year, Japan withdrew from the International Labour Organisation, and all policies that sought a cooperative stand in international politics were abandoned. With the outbreak of the war in China, conscription withdrew all able-bodied men, thus reducing the number of men in the mines. With the country no longer restrained by the labour laws, the way was open for the reinstitution of women miners. In 1937, the Federation of Coal-mining Industries and the Mutual Aid Association for Coal-miners decided to appeal to the Ministry of Industry for policy changes to counter the shortage of workers.[40] Five items were listed in the appeal, including a lifting of the ban on late-night work and on underground work for women and minors. In August 1939, in an ordinance issued by the Welfare Ministry, concerning legislation on pit work by women (Rules on Relief for Miners, Article 11, clause 2(1)), underground work by women was permitted under certain conditions.[41]

The fact that it took Japan only one year after its withdrawal from the ILO to reverse its labour policies is symbolic of the times. The proportion of women mine workers, which had been decreasing steadily, hit a low of 8.3 per cent in 1937, and then rose to 10.6 per cent in 1940 and 18.5 per cent in 1945.[42] Here again we must take note of regional differences. Hokkaido had countered the labour shortage by hiring Koreans, and hence there was little need to hire women: only 3.7 per cent of its workforce in 1940 were women. In Fukushima, Ibaraki, and Yamaguchi prefectures, where the small- and medium-sized mines were concentrated, 15–20 per cent of the workers were women in 1940.[43]

Thus the coal mines under the wartime economy, through the recruitment of women and of labourers forcibly brought to Japan from Korea, managed to increase production. In spite of this, labour productivity declined. During the war, the domestic machinery industry had to produce munitions. This factor limited the further development of coal-mining technology, which had been making progress since the late 1920s. Between 1932 and 1937, one miner extracted 200 tons of coal; in 1940, the figure fell to 174 tons; and in 1944, it dropped to 126 tons. Technological advance came to a standstill, but the wartime economy managed to sustain itself by depending on the large influx of cheap labour from women and Koreans.

Notes

1. Sumiya Mikio, *Nihon sekitan sangyō bunseki* [An Analysis of the Japanese Coal-mining Industry](Iwanami Shoten, Tokyo, 1968), p. 163.
2. *Nihon kōgyō hattatsushi* [A History of the Development of Japanese Mining], vol. 2 (Kōzan Konwakai, 1932), p. 332.

3. Ogino Yoshihiro, "Sangyō kakumeiki ni okeru Chikuhō tankōgyō no rōshi kankei" [Labour Management Relations at Chikuhō mines during the Industrial Revolution], part 3, *Sangyō keizai kenkyū*, vol. 21, no. 1 (1980): 29–30.
4. Hashimoto Tetsuya, "1900–1910-nendai no Miike tankō: sekitan sangyō no sangyō shihon kakuritsu o megutte" [Miike Mine, 1900–1910: The Coal Industry and the Establishment of Industrial Capital], *Mitsui Bunko Ronsō*, vol. 5 (1971): 45–47.
5. Ogino, "Sangyō kakumeiki ni okeru Chikuhō tankōgyō," p. 31.
6. Sumiya, *Nihon sekitan sangyō bunseki*, p. 318.
7. Ibid., p. 320.
8. Yamamoto Sahei, "Chikuhō tankō monogatari" [The Tale of Chikuhō Mine], in *Chikuhō tanko emaki* [Picture Scroll of Chikuhō Mine] (Ashi Shobō, Tokyo, 1973), p. 29.
9. Ministry of Agriculture and Commerce, Bureau of Mines, *Kōfu taigū jirei* [Rules for Treatment of Miners] (1908), p. 40.
10. See Ogino Yoshihiro, "Nihon shihonshugi kakuritsu ki ni okeru tankō rōshi kankei no ni ruikei" [Two Patterns in Labour Management Relations in Mines during the Era of the Establishment of Japanese Capitalism], *Enerugii shi kenkyu noto*, vol. 10 (1979).
11. Housing was provided by the mine operators, hence the low proportion of total expenses spent on housing (4–5 per cent).
12. Uno Kōzō, comp., *Kōza teikokushugi no kenkyū* [Library of Studies on Imperialism], vol. 6: *Nihon shihonshugi* [Japanese Capitalism] (Aoki Shoten, Tokyo, 1973), p.151.
13. *Nihon kōgyō hattatsushi*, vol. 2, pp. 1923-93.
14. Uno, *Kōza teikokushugi no kenkyū*, vol. 6, p. 149.
15. See Tanaka Naoki and Ogino Yoshihiro, "Hogo Kōfu mondai to saitan kikō no gōrika" [Protection of Miners and the Rationalization of Coal-mining], *Nihon Daigaku Seisan Kōgakubu Kenkyū Hōkoku*, vol 11, no. 1 (1978).
16. Ministry of Labour, *Rōdō gyōseishi* [History of Labour Administration], vol. 1 (1961), p. 286.
17. Ibid., p. 287.
18. Sumiya, *Nihon sekitan sangyō bunseki*, p. 386.
19. *Nihon kōgyō hattatsushi*, vol. 2, p. 262.
20. Sumiya, *Nihon sekitan sangyō bunseki*, p. 895.
21. See *Nihon kōgyō hattatsushi*, vol. 2, tables 148–149.
22. The authority on this is *Honpo kōgyō no susei* [Trends in Mining in Japan].
23. Information on the transition to self-sufficiency in mining machinery from *Honpō kōgyō hattatsushī.*
24. Ichihara Ryōhei and Tanaka Mitsuo, "Tankō naya seido no hōkai (2)" [The Collapse of the Naya System in the Mines, part 2], *Nihon rōdō kyōkai zasshi*, 64 (1964); 30–32.
25. Tanaka Naoki, "Chikuhō sekitan kōgyō hattatsu shi gaiyō" [An Outline of the History of Chikuho Sekitan Kōgyō], *Asō Hyakunen-shi* [A Hundred-year History of Asō] (1975), p. 87.
26. Nagasawa Kazuo, "Wagakuni sekitansan ni okeru rōdō kanri no sūsei ni tsuite" [Trends in Labour Management in Japanese Mines], *Nihon kōgyō kaishi*, 544 (1930):713.
27. The population of Korean miners in Fukuoka Prefecture by far outnumbered that in Hokkaido, which had the second-largest group; in Fukuoka, the proportion rose to 44.5 per cent in 1920, 49.5 per cent in 1925, and 62.1 per cent in

1930. From Park Kyung-sik, ed., *Zainichi Chōsenjin kankei shiryōshūsei* [A Collection of Documents on Korean Residents of Japan], vols. 1–2 (San'ichi Shobō, 1975).
28. Miwa Mitsuaki, Noda Nobuo, and Miura Chōhei, *Namazuta tankō kengaku hōkoku* (typed edition, 1926), pp. 41, 46.
29. *Miike Kōgyo enkakushi* [A History of the Development of Miike Kōgyōsho], vol. 7, section 1 on labour, Mitsui Bunko, p. 191.
30. Koga Ryōichi et al., eds., *Kita Kyūshū chihō shakai undō shi nenpyō* [A Chronology of Social Movements in Northern Kyushu] (Nishi Nihon Shuppansha, 1980).
31. Ibid.
32. *Miike Kōgyō enkakushi*, vol. 7, section 1 on labour, p. 191.
33. *Chikuhō sekitan kōgyō shi nenpyō* [A Chronology of Chikuho Sekitan Kogyo] (Nishi Nihon Bunka Kyokai, 1973).
34. Koga, *Kita Kyūshū chihō shakai undō shi nenpyō*.
35. *Chikuhō sekitan kōgyō shi nenpyō*.
36. Ibid.
37. At Miike mines, besides the subsidiary jobs promoted by mine operators, many laid-off workers chose to become vendors of vegetables, fruit, and fish, and they were numerous enough to threaten the business of shops in Ōmuta (Koga, *Kita Kyūshū chihō shakai undō shi nenpyō*).
38. Takahashi Kamekichi, *Nihon sangyō rōdō ron* [Discourse on Industrial Labour in Japan] (Chikuma Shobō, Tokyo, 1937), p. 132.
39. Sangyo Fukuri Kyokai, *Kōjō kōzan no fukuri shisetsu chōsa* (Survey on Welfare Facilities in Factories and Mines) no. 1, Educational and Training Facilities (1933), table 1; Hazama Hiroshi, *Nihon rōmu kanri shi kenkyū* [Studies on the History of Japanese Labour Management] Tokyo, 1 (Daiyamondosha, Tokyo, 1964), fig. 15.
40. Kuboyama Yūzō, *Sekitan kōgyō hattatsushi* [History of the Development of the Coal-mining Industry] (Kōronsha, Tokyo, 1942), p. 166.
41. Ministry of Labour, *Rōdō gyōsei shi* [History of Labour Administration], vol. 1, p. 654.
42. *Honpō kōgyō no sūsei.*
43. In September 1942, a small mine in the Joban mining region had 106 miners, of which 59 were male, 47 women (44 per cent of the workforce); 40 (38 per cent) worked as haulers (Yanase Tetsuya, *Waga kuni chōshō tankōgyō no jūzoku keitai* [The Subordinate State of Japan's Small- and Medium-sized Mines] (Itō Shoten, 1944), pp. 76–77).

Chapter——3

Female Workers of the Urban Lower Class

Akimasa Miyake

This chapter will examine the state of urban lower-class female workers mainly in the period between 1870 and 1920. Although earlier academic works have not focused specifically on female labour as such, much research has been conducted on the pre-war urban lower class. Tsuda Masumi's in-depth 1911–1912 survey of "indigents" (*saimin*) emphasized the proletarian nature of pre-war Japanese society at large.[1] Sumiya Mikio saw the stratum of urban lower class employed in miscellaneous occupations as a pool of surplus labour and the basis for the supply of wage labour.[2] Hyōdō Tsutomu traced the process by which large-scale factory workers rose out of the urban lower class between the 1880s and the post-First World War period.[3]

While the conclusions drawn in these studies vary, all their authors address the urban lower-class issue only in terms of the formation of the wage labour stratum. Regardless of whether the emphasis is placed on the decline into the lower class or the rise out of it, virtually all interest is directed at the formation of the "wage labour" sector. This approach seems to be justified, considering that the development of capitalism leads to class stratification and necessitates the creation of wage labour.

However, while this approach is acceptable from a theoretical standpoint, in reality the urban lower-class condition was not wholly a reduced state from which wage labour attempted to escape. The urban lower class underwent internal change while concomitantly developing (or perhaps retrogressing) independently during the development of capitalism.

Research by Nakagawa Kiyoshi develops along comparable lines. He approaches the pre-Second World War urban lower class, especially that in the inter-war period, as a dynamic element of urban society, and he delineates the changes in the living conditions of this group.[4]

This chapter, which owes much to Nakagawa's research, focuses on the role of female labour as a part of the urban lower class during the 50 years between 1870 and the 1910s.[5]

I. 1870 to the 1880s

1. Historical Antecedents

The historical antecedents to the development of the urban lower class during the rise of capitalism can be found in the status of the *museki musan no kyūmin* (the destitute who possessed neither family registers nor assets of any kind) of Edo at the end of the feudal period. By the eighteenth and nineteenth centuries, the city of Edo had become a huge urban centre of one million people. Unfortunately, few demographic records on the living areas, living standards, occupations, and family structure of the destitute remain, and existing fragmentary references must be pieced together.

The *Fujiokaya nikki*[6] records the living areas of 74,000 "destitute persons" to whom the town councils sold rice at exceptionally low prices in 1860 in an effort to defuse the upsurge of rice uprisings. According to this source, the destitute lived in a wide area encompassing Yotsuya, Shitaya, Shiba, Fukagawa, Honjō, Koishikawa, Kanda, and Hongō, including Yotsuya Samegahashi, Shitaya Mannenchō, and Shiba Shin'ami. In the 1890s, the last three areas became notorious as Tokyo's three largest slums.[7]

As seen below, from the seventeenth and eighteenth centuries these areas had become the ghettoes of the poor. Sources note that many beggars gathered in the lower-lying areas of the Yotsuya Samegahashi, a former swamp site which had been reclaimed in the early Edo period. Shiba Shin'ami first became a settlement for beggars and lowly entertainers in the late seventeenth century. Shitaya Mannenchō, which was formerly called Yamazakichō, gradually grew into a lower-class community from the tenement houses built by masterless samurai Yamamoto Nidayū for ragpickers and other lower-class individuals.[8]

The development of the commodity economy in the late eighteenth and early nineteenth century accelerated the breakdown of the agricultural stratum, and resulted in a build-up of penniless people and destitute farmers who were forced to leave their villages. Most headed for large cities such as Osaka and Edo and settled in the ghettos noted above.

Using the 1828 *Machikata kakiage* [Records of the Commoner Population], Matsumoto Shirō makes the following points about the lower classes' living conditions.[9] First he categorizes the types of households noted in *Machikata kakiage* into those of home-owners, landowners, landlords, property tenants, house tenants and *akidana* (vacant shops). If house tenants are considered the standard for determining the dividing line for the lower class, then a high proportion of the population fell into the lower class, as this sector accounted for over 60 per cent, on average, of all households in each ward of Edo. "Destitute persons" must be considered only a small proportion of the lower class.

As for actual living space, a popular contemporary phrase expresses it simply—"the back alley tenements measured 9 *shaku* 2 *ken*," which is equal to approximately ten square metres. Naturally all well and lavatory facilities were shared. These tenements were located in dark and dreary back alleys, and residents had to make the best of the cramped space and dearth of sunlight.

While the descriptions above shed some light on the living area and living conditions of the lower class, it is more difficult to determine the jobs performed by them. Matsumoto investigates this problem using the slightly later historical records dating from the period when the newly founded Meiji government and big merchants planned to banish the lower-class elements involved in the rice uprisings to the wilderness Shimōsa Plain in Chiba, under the rubric of a "pioneering conquest." Although fragmentary, this source provides an insight into the occupations of the lower class, as seen in table 3.1. On the basis of this table, the lower class can be categorized into four main groups.

- Physical labourers: This group made up the largest proportion of workers of all four categories. The majority were cart-pullers, cart porters and day labourers. Cart-pullers and cart porters were involved in the transportation of commodities, while day labour generally referred to all forms of physical labour.
- Craftsmen: This category included those who worked at customers' residences, such as plasterers and carpenters, and those who worked in their own workshops, such as tatami mat makers and cask-makers. Although the fully-fledged dissolution and decline of this sector did not occur until the latter half of the first decade of the 1900s, many such craftsmen were reduced to the lower class earlier,[10] since these types of professions were easily affected by changing economic conditions.
- Small-scale merchants, fishermen, grocers: Fishmongers and grocers were primarily itinerant vendors. According to Yoshida Kyōichi,[11] who uses different historical references from Matsumoto to study the lower class of the eighteenth and early nineteenth centuries, many women worked by selling anise and flowers (for religious offerings).
- Farmers: Proportionally, farmers made up a small percentage of the lower class. They worked extremely small fields on the outskirts of Edo.

While not included in table 3.1, other historical sources cite a fifth occupational category, or miscellaneous occupations. This group was made up of a jumble of occupations—from itinerant entertainers, diviners, masseurs, and theatre ticket takers to beggars.[12] Women were prevalent in such side work as clothes-washing and errand-running.

No historical references remain to help us ascertain whether the members of the lower class formed families or not. Probably they did, to some extent, especially craftsmen, and the employment of wife and children must have been the norm.

A new breed of lower-class poor appeared immediately after the Meiji Restoration (1868). Large numbers of lower-class samurai, who had been stripped of their privileges, were reduced to poverty. The national and prefectural governments instituted various measures aimed at aiding this group of ruined samurai. In 1869, the Tokyo Metropolitan Government estab-

Table 3.1a. Former Occupations and Residences of Farmers of Niwamura, 1869

Occupation	No. of former households	Previous domicile
1. Cart-drawer	12	Mikawachō 2, Asakusa Shibazakichō, Shitaya Kurumazakachō, Mikasachō 1-chōme, Kikuzakadaichō, Abekawachō, Ushigome Kaitaichō, Ushigome Gokenchō, Koishikawa Haramachi, Samegahashi Tanimachi, Samegahashi Minamimachi
Cart porter	7	Mikawachō 3-chōme, Yoshiokachō 1-chōme, Shitaya Yamabukichō, Fukagawa Reiganchō, Hongō Motomachi, Koishi-kawa Sakashitachō, Shitaya Inarichō
Day labourer	4	Honjo Hanachō, Yushima Ryōmonzen-chō, Ushigome Konandochō, Samega-hashi Tanimachi
Palanquin-bearer	4	Kanda Nakamachi 3-chōme, Mikasachō 1-chōme, Shitaya Shinmachi, Komagome Asagachō
Fireman	2	Mikawachō 4-chōme, Hongō Higashi Takemachi
Subtotal	29	
2. Plasterer	2	Shitaya Yamabushichō, Ushigome Bentenchō
Carpenter	2	Shitaya Torishinmachi, Samegahashi Minamichō
Tatami-maker	2	Kanda Suehirochō, Nakanogōhashichō
Dyer	2	Akasaka Sojimachi, Kikuzaka Daimachi
Well-digger	1	Nishinokubo Hirōcho
Sawyer	1	Asakusa San'yachō
Cask-maker	1	Shitaya Mannenchō
Clog-maker	1	Azabu Yamamotochō
Pillow cover maker	1	Fukagawa Kagachō
Paper decoration maker	1	Kōjimachi Hirakawachō
Metal-caster	1	Fukagawa Saruechō
Wooden utensil maker	1	Kyōbashi Gorobechō
Glassmaker	1	Ushigome Tenjinchō
Blacksmith	1	Akasakachō
Dyer's assistant	1	Yanagihara Yanagichō
————	1	Honjo Kitashinmachi
Coin-minter	1	Asakusa Motoyoshichō
Tile-maker	1	Ikenohata Shichikenchō
Subtotal	22	

Table 3.1a (*continued*)

Occupation	No. of former households	Previous domicile
3. Fish-seller	7	Koishikawa Sakaichō 2, Kanada Aioichō, Kanda Sue-hirochō, Shitaya Yamabukichō, Asakusa Tajimachō, Samegaha-shi Omotemachi
Vegetable-seller	5	Kanda Aioichō, Shitaya Torishin Machi, Shitaya Yamabukichō, Asakusa Tajimachō, Samegahashi Hachikenchō
Furniture-dealer	2	Shiba Shin'amichō, Kanda Suehirochō
Kimono-dealer	1	Kyōbashi Gorobechō
Paper-dealer	1	Asakusa Tajimachō
Rice-seller	1	Shimō Tomizakachō
Tabi-seller	1	Asakusa Shibazakichō
Charcoal-seller	1	Shiba Shin'amichō
Confectioner	1	Shibatachō
Kitchen ware	1	Asakusa Hatagochō
Charcoal-dealer	1	Kanda Sakumachō
Barkeeper	1	Shitaya Minowachō
Cook	1	Shitaya Mannenchō
4. Farmer	3	Asakusa Fukuichō, Asakusachō, Nakanogō Azumachō
5. Undetermined	1	Kanda Matsushimachō
Total	79	

Source: Compiled from Matsumoto Shirō, "Bakumatsu, ishin-ki ni okeru toshi no kōzō" [Urban Infrastructure during the Late Feudal and Early Meiji Periods], *Mitsui bunko ronsō*, no. 4 (1970), table 6.

lished two relief centres where the family members of ex-samurai were given occupational training. The children were taught weaving, as well as techniques for making silk thread, fabric clog thongs, Indian ink, and paper.

The governor of Tokyo also called together the city's influential businessmen to form a building and repair chamber of commerce. This organization opened a workhouse in 1873. As of 1876, the workhouse came under the direct control of the Tokyo government and began the manufacture of leather shoes, paper, and matchboxes.

An institution for the poor was also established by the Osaka government in 1871. The next year it was reorganized into an occupational training centre, where weaving, papermaking, and spinning were taught. The Kyoto government also set up a similar establishment in 1870. Here too, oil-

Table 3.1b. Former Occupations and Residences of Farmers from Hatsutomimura, 1872

Occupation	No. of former households	Previous domicile
1. Day labourer	17	Shitaya Mannenchō 2-chōme 5, Shitaya Oyosumichō 4, Azabu Ryōdochō 4, Ikenohata Shichikenchō 2, Shitaya Mannenchō 1-chōme, Shitaya Kurumazakachō
Cart-pullers	5	Azabu Ryōdochō, Shitaya Mannenchō 1-chōme, Shitaya Kurumazakachō, Kanda Aioichō, Ushigome Wakamatsuchō
Subtotal	22	
2. Carpenter	3	Shitaya Toyosumichō, Kanda Shinshirogane, Nezu Miyanagachō
Gardener	3	Shitaya Toyosumichō 2, Shitaya Kurumazakachō
Plasterer	2	Shitaya Mannenchō 2-chōme, Ushigome Wakamatsuchō
Bucket-maker	2	Azabu Roppongichō, Shitaya Mannenchō 2-chōme
Palanquin-bearer	1	Shitaya Mannenchō 1-chōme
Plasterer's helper	1	Shitaya Mannenchō 2-chōme
Wooden utensil maker	1	Shitaya Mannenchō 2-chōme
Blacksmith	1	Shitaya Toyosumichō
Subtotal	14	
3. Greengrocer	5	Shitaya Mannenchō 2-chōme 2, Shitaya Kurumazakachō, Shitaya Toyosumicho, Fukagawa Yanagawachō
Fish-seller	3	Shitaya Mannenchō 2-chōme, Shitaya Toyosumichō, Shitaya Mannenchō 1-chōme
Seasonal goods seller	2	Shitaya Toyosumichō, Ikenohata Shichikenchō
Furniture	2	Shitaya Toyosumichō, Kanda Suehiro-chō
Second-hand goods	1	Azabu Ryōdochō
Wigs/hairpieces	1	Shitaya Mannenchō 2-chōme
Vegetable-seller	1	Shitaya Toyosumichō
Subtotal	15	
4. Farmer	2	Ushigome Wakamatsuchō 2
5. Undetermined	2	
Total	55	

Source: Compiled from Matsumoto Shirō, "Bakumatsu, ishin-ki ni okeru toshi no kōzō" [Urban Infrastructure during the Late Feudal and Early Meiji Periods], *Mitsui bunko ronsō*, no. 4 (1970), table 6.

pressing, candle-making, paper manufacture, weaving, and comparable skills were taught. In 1875, an establishment was inaugurated in Kyoto in which the destitute were paid allowances for collecting garbage.[13]

In 1874 the national government promulgated a relief regulation which enlisted the magnanimity of the general population in assisting the lower class, while minimizing national aid as much as possible. As a result, the policies noted above, instituted at the prefectural and metropolitan level, represented the relief measures in real terms. In 1882–1883, however, these policies were either abolished or left to be implemented by private organizations. Commoners as a whole became much poorer as a result of Finance Minister Matsukata's deflationary policies, so that providing relief to only a part of this class was meaningless.[14]

2. The Expansion of the Lower Class

The so-called "primitive accumulation" stage of capitalism in Japan (in which labour supply, monetary resources, and production methods had accumulated to levels sufficient to form the basis of capitalistic development) can be dated to the period between the Matsukata deflationary policies and the earlier part of the 1890s. It has been said that, unlike in Western Europe, where large masses of poor suddenly appeared in the urban centres, the numbers of persons displaced from the countryside to the cities were relatively small during this period of primitive accumulation.

In recent research, Unno Fukuju contests this theory. He insists that labour statistics point to a large inflow of farmers into the cities. In fact, of the 920,000 persons residing in cheap lodging-houses in Tokyo in 1884, only 44 per cent were registered as Tokyo residents, while the remainder had come from other areas throughout the nation.[15]

Historical records state that the 10,000 residents of the Nagomachi area of Osaka, one of Japan's four largest slums (the other three were Yotsuya Samegahashi, Shitaya Mannenchō, and Shiba Shin'ami), were divided into two groups: the original residents, who had fallen into dire circumstances, and the newly settled residents.[16] There was also a considerable increase in the inflow of seasonal labour into Tokyo at this time.[17]

Owing to the rapidly growing numbers of the poverty-stricken, the settlements for the poor began to expand from the original ghettoes to the surrounding areas. The residential status of the expanded lower class is not thought to have changed from earlier patterns.[18] In this way, the urban lower class became inflated through the "primitive accumulation" process.

Eventually the rapid growth of the lower class attracted wide social attention. Surveys on slum conditions were taken up in several newspapers. Of these, the following two mentioned the state of women workers in depth.

Kure Ayatoshi's survey on the status of Tokyo's destitute, published in volume 57 (1891) of *Sutachisuchikku zasshi* [Statistics Magazine],[19] was "a collection of excerpts taken from a certain newspaper's survey of the state of Tokyo's poor during the escalation of rice prices last year." In it, the occupations of men and women from each ward of Tokyo ware noted as follows.

> Day labourers, construction workers, jinriki-pullers, merchants at temple fairs, low-grade factory workers, used paper and junk collectors, clog-repairers, dancers, outdoor storytellers, outdoor exhibitioners, fishermen, shell fishermen, bait-diggers, peddlers, papermakers, makers of paper ornaments for hair and decorative uses, fishmongers, and vegetable-sellers. For women and children, cart-pushing and itinerant tea-picking.

While "low-grade" factory workers are not discussed in detail, this is taken up in Matsuhara Iwagoro's 1893 work entitled *Tōkyō shinai no shukōjin katō shakai seigyō no ichidai genshō* [The Serious Situation of Lower-level Manual Labourers of Tokyo].[20] These people could be considered as the manual labour sector of the miscellaneous occupations group. For example, many women and children worked in factories which manufactured umbrella ribs in what was once side work but developed into a primary source of work. In the Honjō area of Tokyo alone, there were 8,000 of these manual labourers, and of them two-thirds were women and children. During this period, most lower-class women workers did not find themselves in circumstances greatly different from those of the late Edo period. However, the appearance of new types of low-grade work merits attention.

Suzuki Umeshirō's 1888 survey of the Nagomachi slum in Osaka, *Osaka Nagomachi hinminkutsu shisatsuki*, published in 1918,[21] gives a specific breakdown of the occupations of those living in Nagomachi. These occupations have been categorized in table 3.2. The following points concerning female labour can be drawn from the table. First, one of the first industries to appear during the early period of capitalism was match-manufacturing, which employed a relatively large number of workers. In sharp contrast to the small percentage of men, over 10 per cent of all women were engaged in this work. The actual working conditions in the match-manufacturing industry will be taken up in the next section.

Secondly, in terms of occupation, the largest proportion of persons (16.4 per cent of all workers) were involved in garbage-collection. With the inclusion of miscellaneous jobs, such as old clog collection and sales of used casks (which fall in the category of "miscellaneous jobs" in the table), the percentage of persons involved in the collection of discarded objects amounted to close to 30 per cent.[22] Both men and women were involved in this work in approximately equal proportions.

Thirdly, approximately 26 per cent of the women were unemployed. While it is natural for girls under 15 to be unemployed, it is notable that a fair proportion those over 15 were not employed. It is doubtful, however, that these women were entirely without work. Suzuki notes that theft was another type of occupation among the poor of Nagomachi. Taking into account this type of "occupation," the numbers of so-called unemployed are reduced by a large margin.

Table 3.2. Work Performed by the Poor of Nagomachi, 1888

Jobs	Females (A)			Males (B)					
	<15	>15	Total	<15	>15	Total	A + B	Females (%)	Males (%)
Garbage collection	261	433	694	234	164	398	1,092	16.4	9.2
Match production	257	211	468	122	51	173	641	11.1	4.0
Miscellaneous[a]	54	352	406	72	628	700	1,106	9.6	16.3
"Employment"[b]	102	203	305	143	374	517	822	7.2	12.0
Umbrella-making	68	175	243	79	271	350	593	5.8	8.1
Begging	123	112	235	159	87	246	481	5.6	5.7
Merchant trades	66	165	228	80	169	249	477	5.4	5.8
Factory work[c]	25	115	140	26	226	252	392	3.3	5.9
Sweet-making	26	58	84	6	56	62	146	2.0	1.4
Pawning	15	49	64	21	87	108	172	1.5	2.5
Junk collection	0	63	63	3	39	42	105	1.5	1.0
Entertainment[d]	18	43	61	10	64	74	135	1.4	1.7
Restaurant	14	34	48	7	37	44	92	1.1	1.0
Rental business	11	28	39	12	39	51	90	0.9	1.2
Carter	0	0	0	0	330	330	330	0	7.7
Student	54	0	54	88	0	88	142	1.3	2.0
Unemployed	709	384	1,093	556	67	623	1,716	25.9	14.5
Total	1,803	2,422	4,225	1,618	2,689	4,307	8,532	100.0	100.0

a. Matchbox assembly, tobacco pipe replacement, old clog collection, used cask sales, purchase of broken lamp glass, dog-killers, cat-catchers, river-gleaners (for needles, nails and gold, silver or copper coins), recyclable rubbish gatherers (for waste paper, rag fragments, metal objects, etc.), migrant pilgrims, religious worshippers, etc.

b. Persons without fixed occupations.

c. Nagomachi fan-makers, hemp rope makers, weavers of hemp soled sandals, etc.

d. Performers of *shinnei*, *gidayū*, *saemon*, *hitotsutoseibushi*, and *ukarebushi* (shamisen genres), women sumo wrestlers, magicians, storytellers, dancers, Western-style magicians, etc.

Source: Suzuki Umeshirō, *Osaka nagomachi hinminkutsū shisatsuki* [Account of Observations of Osaka's Nagomachi Slum], reproduced in *Meiji zenki no toshikasō shakai* [The Urban Lower Class in the Early Meiji Period], annotated by Nishida Taketoshi (Kōseikan, Tokyo, 1970).

Based on the records from Nagomachi and Tokyo, it is possible to categorize as follows urban lower-class female labour in the 1870s and 1880s.

- Physical labour: This category applies to tea-pickers or cart-pushers in Tokyo, and day labourers in Nagomachi. As in the late feudal period, this group encompassed many different kinds of people engaged in physical labour.
- Craftsmen: this included fan-makers, rope-makers, rope-weavers, and other labour involved in manual production. It is likely that most of these labourers did such work as a side job at home.
- Miscellaneous occupations: This included such work as rubbish-collecting, entertainment, and petty trading operations.
- Factory work: The leading type of factory work was related to match-manufacturing, one of the first occupations to utilize lower-class women during the process of capital formation.

These four categories were in no way fixed in terms of substance or classification. While the respective occupations had a certain degree of independence from each other, there was also a degree of fluidity among the occupations—an entertainer could turn to rubbish-collecting, for example.[23] During this time, occupational trends turned away from all other categories toward factory work. Little is known about the state of lower-class families during the 1870s and 1880s. Judging from the age brackets found in table 3.2, it is certain that lower-class workers had families. However, since the head of the household and the spouse had extremely meagre incomes, it is likely that all family members were employed in some capacity.

II. The Industrial Revolution

1. The Development of the Urban Lower Class

Japan's industrial revolution began in 1886–1889, with the sudden rise of private enterprise, and went on to 1907.[24] As shown in tables 3.3 and 3.4, Tokyo's slums expanded greatly during this period. Table 3.4 shows the number of cheap lodgings in each ward. These cheap lodgings housed the displaced rural folk who flocked from throughout the country to the cities in search of work. The tables demonstrate how the areas where the lower class congregated differed considerably before and after 1896. From the period of "primitive accumulation" to the early industrial revolution period, most lived in the Asakusa-Shitaya areas, whereas from the mid to late industrial revolution era, the highest concentrations of poor were in the Fukagawa and Honjō areas, which were Tokyo's industrial centres at the time. This trend reflects the growing numbers of the lower class involved in factory work.

Table 3.5 shows the numbers of private factories in Tokyo from 1889 to 1907. From this period to the early industrial revolution period, the facto-

Table 3.3. Appearance of New Slums in Tokyo

	1877–86	1887–96	1897–1906	1907–12	1913–23
Fukagawa ward	0	2	2	4	2
Honjō ward	1	1	3	8	2
Asakusa ward	1	7	5	3	1
Shitaya ward	1	7	2	2	0
Other wards	8	5	2	0	15
Total	11	22	14	17	20

Source: Tsuda Masumi, *Nihon no toshi kasō* [Japan's Urban Underclass] (Keisō Shobō, Tokyo, 1985).

Table 3.4. Numbers of Cheap Lodgings in Tokyo, by Ward

	1897	1902	1907	1912	1916
Fukagawa ward	46	83	112	105	118
Honjō ward	78	110	130	150	154
Asakusa ward	32	45	51	57	75
Shiba ward	3	3	3	2	1
Azabu ward	3	10	18	20	18
Yotsuya ward	16	17	27	26	25
Hongō ward	3	3	3	3	3
Total	181	271	344	363	394

Source: Tokyo-shi Shakaikyoku [Tokyo Municipal Bureau of Social Affairs], ed., *Tōkyō shinai no kichin'yado ni kansuru chōsa* [Survey on Cheap Lodgings in Tokyo] (1923).

ries were primarily involved in the production of matchboxes, rubber, and soap, as well as various types of chemical industries. As can be seen from the table, textile and heavy industry factories increased gradually. Osaka, which was then Japan's largest industrial centre, saw a comparable trend.[25]

Table 3.6 tabulates the changes in the number of female workers in Tokyo, by occupation. Both tables 3.5 and 3.6 clearly reflect the high proportion of labour in the chemical industries (over 90 per cent were involved in the match industry) at the start of the industrial revolution, and the shift to the textile industry, particularly the spinning industry, in the 1900s. (The reason for the low percentage of workers in table 3.6 in the heavy industries, which had many factories in the latter part of the industrial revolution—see table 3.5—is the low percentage of female labour in these fields.)

All these industries turned to urban lower-class women as a source of

Table 3.5. Changes in the Number of Private Factories in Tokyo

	1889	1893	1896	1901	1907
Textiles					
Silk thread	2	13	46	23	27
Spinning	7	2	2	15	10
Weaving	3	7	43	64	102
Subtotal	12	22	91	102	139
Machinery, appliances, metal industries					
Machinery, appliances	21	22	57	77	160
Metal-processing	4	6	7	16	33
Shipbuilding	1	2	3	7	6
Subtotal	26	30	67	100	199
Chemicals[a]	27	46	58	45	78
Leather industry	5	8	8	10	16
Printing, bookbinding	32	16	33	50	82
Papermaking	0	10	11	9	9
Food-processing	10	10	33	49	59
Gas, electricity	0	5	3	9	11
Ceramics	14	26	11	25	42
Household goods	8	16	13	21	60
Miscellaneous	5	3	6	32	70
Total	139	192	334	452	765

a. Primarily the manufacture of household goods such as matches, rubber products, and soap.

Source: Ishizuka Hiromichi, *Tōkyō no shakai keizaishi*, p. 93 (based on *Tōkyō-fu tōkeishō* [Statistical Records on Tokyo], 1889–1907).

labour.[26] The discussion below considers the match industry as a case-study.[27]

2. Female Labour in the Match Industry

Match-manufacturing technology, which was introduced to Japan during the 1870s, quickly became established due to the simplicity of the process, and soon became one of Japan's leading export industries. According to a study by Takamura Naosuke on the industrial and trade structure during the industrial revolution,[28] from 1898 to 1900 matches accounted for 3.1 per cent of Japan's average annual export volume, and were ranked fourth behind raw silk thread, cotton thread, and silk fabric (*habutae*). By the latter part of the industrial revolution, this figure had shrunk to 2.5 per cent, and by 1910 matches were ranked seventh or eighth among Japan's major exports.[29] As

Table 3.6. Fluctuations in Female Occupations in Tokyo

	1882	1887	1894	1900	1904	1910
Textiles						
Silk thread	40	40	1,374	1,321	1,576	2,214
Spinning	0	24	1,851	4,375	6,405	10,191
Weaving	186	65	1,396	4,241	5,321	15,492
Subtotal	226	129	4,621	9,937	13,302	28,897
Machinery, appliances, metal industries						
Machinery, appliances	0	3	129	117	220	557
Metal-processing	0	0	0	13	76	507
Shipbuilding	0	0	0	9	0	16
Subtotal	0	3	129	139	296	1,080
Chemical industries	969	699	1,424	462	599	1,118
(incl. match maufacturing)	941	654	1,313	314[a]	219	191
Leather industry	12	0	265	12	1,009	63
Printing, bookbinding	0	55	181	528	831	1,238
Papermaking	249	319	289	407	312	379
Food-processing	0	0	2,927	1,392	2,061	1,283
Gas, electricity	0	0	0	0	0	3
Ceramics	0	83	150	211	154	407
Household goods/ miscellaneous occupations	37	114	290	141	754	1,765
Printing Bureau factory	0	1,036	559	1,006	1,554	1,428
Senjū Woollen Manufacturing	0	245	418	495	1,282	0
Tokyo Artillery Factory	0	0	0	334	3,950[b]	1,871
Govt Monopolies Bureau	0	0	0	0	0	7,294

a. Of whom 43 were under 14 years old.
b. Of whom 2,478 were involved in the manufacture of firearms.
Source: *Tōkyō-fu tōkeisho* [Statistical Records on Tokyo] (1882–1910).

seen from table 3.7, over 80 per cent of the matches produced were for foreign consumption. In the early phases of match production, the main importing country was China; later, importing countries ranged across South-East and South Asia.

By the industrial revolution, matches were produced throughout all regions in Japan, although in 1895 90 per cent of production was concentrated in Hyōgo, Osaka, Aichi, and Tokyo. As for the concentration of factories, in 1899 the above-mentioned four regions accounted for 70 per cent of all factories (with more than 10 employees) nationwide, with 62 factories in Hyōgo, 35 in Osaka, 12 in Aichi, and 17 in Tokyo. These four regions also accounted for close to 80 per cent of all the employees in the match-manufacturing industry.[30] Hyōgo Prefecture and Osaka boasted the highest

Table 3.7. Match Factory Indices

Year	Production volume	Percentage exported	No. of factories	Workers			Per capita productivity
				Male	Female	Total	
1898	22,226,289	99	264	5,442	14,466	19,908	1,116.5
1899	25,647,725	77	278	5,203	14,026	19,229	1,333.8
1900	21,354,801	90	289	5,228	12,863	18,091	1,180.4
1901	32,901,319	76	261	5,656	16,504	22,160	1,484.7
1902	27,400,508	100	244	4,977	15,064	20,041	1,367.2
1903	32,392,739	88	251	6,294	14,592	20,886	1,550.9
1904	35,301,434	94	219	6,070	15,835	21,905	1,611.6
1905	38,842,947	97	254	5,768	18,761	24,529	1,583.6
1906	54,802,293	70	250	5,468	18,721	24,189	2,265.6
1907	57,125,761	59	257	6,942	16,773	23,715	2,408.8
1908	39,397,680	86	213	4,878	11,828	16,706	2,358.3
1909	49,972,039	83	214	5,288	12,663	17,951	2,783.8
1910	49,947,215	76	203	4,998	12,981	17,979	2,778.1
1911	43,948,327	85	195	4,635	11,742	16,377	2,683.5
1912	52,845,232	85	189	4,560	11,819	16,379	3,226.4
1913	51,731,010	85	189	4,907	11,628	16,535	3,128.6
1914	49,050,229	81	181	4,001	11,663	15,664	3,131.4

Source: Yamashita Naoto, "Keiseiki Nihon shihon shugi ni okeru matchi kōgyō to Mitsui Bussan" [The Match Industry and Mitsui Bussan during the Formative Period of Capitalism in Japan], *Mitsui bunko ronsō*, vol. 6 (1972), tables 6 and 9.

figures in terms of production volume, factory numbers, and numbers of employees. This can be accounted for by the proximity of Kōbe, the leading trading port at the time; this facilitated the transportation of matches, which had become an export commodity.

Match manufacturing was conducted on a medium-sized to tiny scale. According to the 1909 *Statistical Records on Factories*, there were no match factories with over 1,000 employees. Factories employing between 5 and 99 workers accounted for over 70 per cent of all match-manufacturing factories. This situation remained essentially unchanged. The simple nature of the match-manufacturing process accounts for the small scale of the factories and the wide employment of female labour, which formed the majority of urban underclass workers.

The main processes involved in the match-manufacturing process were three: production of the matchsticks, and construction of the matchboxes and of the match head itself. Most match factories were supplied with matchsticks and matchboxes produced elsewhere.[31] Match splint manufacturing plants were initially centred in the Hokkaido and Tohoku areas, owing to the proximity of white poplar forests, which provided the raw material, but in the 1900s they began appearing in the Kōbe-Osaka area, where match manufacturers were concentrated.

As for the matchboxes, generally the match manufacturers would purchase the materials and farm out the assembly work and work of applying the brandname seals to the boxes. This work was usually farmed out to urban lower-class women, who did it at home as piecework. Case-studies from the Kōbe-Osaka area reveal that these women were not directly hired by the match manufacturers, but by the small matchbox-makers working on contract from the manufacturers.

Excluding the matchstick and matchbox production process, the manufacture of matches involved the following ten steps: preparation of the wood splints, alignment of splints, stamping, application of paraffin, application of the chemical composition, drying, packing into boxes, cleaning of the striking surface, final inspection, and packaging. Of these, the most technically important aspect was the alignment of the splints for dipping the match head.

As shown in table 3.7, women accounted for 70 per cent of match-manufacturing factory workers. Most of them were involved in three of the above operations—the alignment of matches, packing, and packaging—as noted in the Ministry of Agriculture and Commerce report on the conditions of factory workers, which states that "the primary job of women in the match-manufacturing industry was preparing the matchsticks (although boys also did this type of work). Packing the matchboxes and wrapping the packages were done solely by women."[32] Female workers were pivotal to the match-manufacturing industry, both quantitatively and qualitatively.

It would be expected that the technological revolution of the matchstick alignment process, the most technically demanding aspect of the match-manufacturing process, would reduce the numbers of women workers. In the early 1890s, a German matchstick-alignment machine was introduced. Because "the German matchstick-alignment machines can only be operated by adult men,"[33] it was assumed that women would be excluded from the alignment process. However, very few factories were able to introduce the German machinery, owing to its high cost. Instead, in the late 1890s, a Japanese-made foot-pedal-operated alignment machine was introduced and its use began to spread. These machines were operated by older women; thus, women were not eliminated from the operations even after automation with the Japanese equipment.[34]

The introduction of this equipment helped raise the productivity of Japanese match factories. As seen in the right-hand column of table 3.7, per capita productivity increased 1.5-fold between 1905 and 1906, a period that coincides precisely with the introduction of the Japanese matchstick-alignment machines.

Regarding the supply of female labour in the match factories, Yokoyama Gennosuke notes on page 126 of *Nihon no kasō shakai*:

> Regarding the relationship between match factories and the lower class, in places such as Kōbe, where maintaining a livelihood is uncertain, the growth of slums can be attributed to multiple causes. In areas

Table 3.8. Case-study of Match-factory Workers and Their Families[a]

	Head of household					Wife			Children					
No.	Sex	Former job	Age	Work	Average daily wage (sen)	Age	Work	Average daily wage (sen)	No.	No. working	Sex	Age	Work	Average daily wage (sen)
1	M		42	Police assistant	23		Rope-twister	10	2	1	M	11	Match-factory worker	6
2	M	Agri-culture	24	Labourer	24	23	Hemp rope twister	18	1	0				
3	M	Agri-culture	58	Servant	?	57			3	1	F	22	Hemp rope twister	8
4	M		49	Craftsman	30	42	Caretaker of abandoned children	10	2	2	① F	15	Match-factory worker	9
											② M	12	Match-factory worker	5
5	M		26	*Mokugyo* (wooden gong in shape of fish) maker	30									
6	M		40	Servant	24	39	Sandpaper seller	10	4	2	① F	18	Match-factory worker	8
											② M	13	Match-factory worker	8
7	M	Cart-puller	46	Sewer cleaner	13	46	Home worker	5	5	3	① F	17	Match-factory worker	13
											② M	14	Match-factory worker	8
											③ M	11	Match-factory worker	6
8	M	Cart-puller	45	Sanitation worker	24	39			4	2	① F	18	Shop apprentice	10
											② M	15	Match-factory worker	8
9	M	Agri-culture	30	Longshore-man	30	25	Home worker (matchbox assembly work)	3	1	0				
10	M			Painter	34		Home worker (matchbox assembly work)	5	1	1	M	16	Longshoreman	17
11	M		51	Hospital janitor	40	42	Home worker (matchbox assembly work)	5	4	3	① F	21	Match-factory worker	8
											② F	19	Match-factory worker	10
											③ F	17	Match-factory worker	10
12	M		46	Cauldron cleaner	28	48	Match-factory worker	6						
13	M	Crafts-man	32	Longshore-man	40	22	Match-factory worker	7						
14	M		40	Undertaker's assistant	30	46			2	1	F	14	Match-factory worker	7
15	M		40	Servant	24	37	Match-factory worker	10	4	1	F	13	Nursemaid	

a. Figures in the income and expenditure columns are calculated on a daily basis.
Source: *Matchi shokkō jijō*, pp. 154–176.

Cohabitants						Breakdown of expenditure (sen)						
No.	Relationship	Age	Work	Average daily wage (sen)	Total income (sen)	Rent	Food expenses	Bedding rental fees	Educational expenses	Medical expenses	Total expenditure	Remarks
					39	4	38		4	16	62	Head of household has eye disease
2	Brother ①	17	Labourer	24	72	4	95	3	7		109	
	Brother ②	12	Match-factory worker	6								
					8 + α	4	44				48	Wife is ill, daughter formerly worked in match factory
					54	4	27		8		39	
6	Mother	51			95	2	53				55	Brother no. 1 has cerebral disease
	Brother ①	24										
	Brother ②	23	Match-factory worker	25								
	Brother ③	19	Match-factory worker	20								
	Brother ④	16	Match-factory worker	12								
	Brother ⑤	14	Match-factory worker	8								
					50	3	30	3			36	
					45	2	45	3			50	
					42	3	32				35	Wife is ill
					33	6	33	6			45	
					56	6	32				38	
					73	6	63		1		70	Eldest daughter is ill
					34	8	23	8			39	
1	Nephew	22	Longshoreman	40	87	8	24				32	
					37	5	28				33	
					34	3	30				33	

> such as Osaka, however, where there are few occupations for the lower class, many people are reduced to poverty and life in the ghettoes. While the change in the appearance of Nagomachi, Japan's leading slum, can in part be attributed to other factors, the leading element behind this change is the establishment of many match factories near the area.

As this commentary suggests, the match factories were first established and underwent development based on a supply of urban lower-class labour, particularly young workers. Moreover, the appearance of the match factories helped to alter the living conditions of the urban lower class. Page 153 of *Matchi shokkō jijō* [Conditions of Match-factory Workers] describes the relation between the match factories and the urban lower class in the following manner:

> Owing to the fact that match-factory workers are generally the poor from the ghettoes, the two are bound by an inseparable tie. The factory workers commute from their homes or from cheap lodgings, and most match factories are located on the outskirts of town, not far from these slum areas. The reason for locating factories in such areas is to ensure a steady supply of factory employees. Typical examples include certain portions of Honjo, Fukagawa, and Asakusa in Tokyo, areas adjacent to Imamiya and Nanba in Osaka, the Tachibanacho-dori, Kawasaki, and Fukiai areas of Kōbe, and the outlying areas of Hyōgo.

Matchi shokkō jijō notes that the best way of determining the living conditions of the match-factory workers is to investigate the extent and types of living expenses incurred by those living in the ghettoes, and provides case examples of the living conditions of 22 urban underclass families in the Osaka area. Of these, 15 families (table 3.8) were connected with the match factories in some capacity. The following four points can be induced from the table:

1. Families with family members employed at the match factories were all multiple-income families. Match-factory workers were generally adult women (mainly wives) or adolescent boys and girls. Most of the family members were forced to work since the income of the head of the household was far from sufficient to raise the entire family.
2. Since several members of each family were generally employed, the wages paid by the match factory were considered supplementary income and, as a result, were extremely low. Those receiving a daily wage of less than 10 sen (one one-hundredth of a yen) greatly outnumbered those with higher incomes. In this connection it should be mentioned that even women employed as power-loom operators, who, according to annual statistics, had the lowest wages among the various job sectors, received approximately 20 sen in daily wages in the first half of the 1900 decade.[35] These low wages of match-factory workers kept their family finances

chronically debt-ridden, as can be seen from examples 1, 2, 7, and 12 in table 3.8. Yokoyama's remark that the emergence of match factories altered the look of the lower class is relative. Even when the match factories were at their height, the conditions of the underclass did not improve markedly.

3. The assembling and labelling of the matchboxes were tasks primarily farmed out to women who did the work at home. Their income was even less than that of the match-factory workers. Generally a worker assembled 2,000 boxes a day. The wage per 1,000 boxes was 7 sen, with 1 sen 5 rin (1 rin is one-tenth of a sen) deducted for paste, leaving 5 sen 5 rin.[36]
4. During the two decades of the 1870s and 1880s, most urban underclass families had a family member employed in either the match-manufacturing factories or matchbox assembly work. As seen in table 3.8, match-factory work gradually came to replace physical labour and other factory work as the primary occupation of the head of the household.

3. Factory Conditions

Now let us look at conditions in the match factories. The following sections will introduce the labour composition, employment structure, consecutive years of labour, and working conditions of the match-factory workers.

(1) Composition of the Labour Force

In this section, we will look at the age structure and functions performed by female labour. The age structure of workers is described in *Matchi shokkō jijō* [Conditions of Match-factory Workers], a report on a study by the Agriculture and Commerce Ministry on 14 medium-sized (from 34 to 1,200 employees) match factories in Osaka between 1900 and 1902 (table 3.9). Minors of under 14 years of age were found by this study to account for 18.9 per cent of the workforce, but this figure is qualified with the statement that: "Although in most factories the minimum hiring age is 12 or 13 years of age, even very young children without any training can earn 2 or 3 sen a day working at a match factory; impoverished children lie about their age to

Table 3.9. Age of Workers in Match Factories, 1900–1902

Age (years)	Males	% of total	Females	% of total	Total	% of total
≤10	62	1.2	132	2.5	194	3.7
10–13	196	3.7	613	11.5	809	15.2
14–19	454	8.5	1,609	30.2	2,063	38.7
≥20	622	11.7	1,642	30.7	2,264	42.4
Total	1,334	25.0	3,996	75.0	5,330	100.0

Source: *Matchi shokkō jijō* [Conditions of Match-factory Workers], pp. 130–131.

Table 3.10. Division of Roles by Age and Sex among Workers in Match Factories, 1902

Job description	Sex	<14	14–19	≥20	Total
Matchstick alignment	Male	—	27	48	75
	Female	3	54	41	98
Composition dipping	Male	10	27	52	89
	Female	—	5	13	18
Filling boxes	Male	6	—	6	12
	Female	132	205	261	598
Painting sides	Male	4	2	18	24
	Female	4	7	10	21
Packaging	Male	—	—	—	—
	Female	5	7	33	45
Labelling	Male	—	—	—	—
	Female	28	19	58	105

Source: *Matchi shokkō jijō* [Conditions of Match-factory Workers], p. 132.

get a job; factories let them work unless circumstances force them to investigate." Thus workers under 14 years of age accounted for a greater percentage than indicated in table 3.9; the cumulative percentage of workers of under 20 years of age indicates that the match-factory workforce was clearly composed primarily of young female workers, that is, girls and young women.

Matchi shokkō jijō gives detailed information on the work they performed in five factories (table 3.10). Of the different stages of match production, box labelling was usually done outside, so the data in the table refer to only three of the five factories studied. Female workers working in the factory were primarily responsible for filling boxes and lining up matchsticks; the majority of workers doing these jobs were under 20.

In short, girls and young women were the mainstay of the match factory workforce, and most were engaged in labour-intensive manual activities (tables 3.9 and 3.10).

(2) Form and Duration of Employment

Match-factory workers "lived for the most part in nearby poor neighbourhoods and none came from very far away; thus no special recruiter had to be sent out to find workers. Usually, workers were recruited by posting signs at the entrance to the factory or on street corners."[37] Unlike heavy industry, where new workers were introduced through the offices of an *oyakata* (boss), or the spinning and silk-reeling industries, where recruiters supplied female labour, the match industry did not actively recruit, but tended to rely on local hiring because of the inherently small scale and the siting conditions of match factories. The simplicity of this hiring approach led to "the absence of contracts specifying periods of employment or advance payment of wages" and frequently "made the work of a match-maker utterly unplanned and irregular, with workers working one day at one factory and the next day at another occupation or at a different match factory."[38] Thus, "needless to say, periods of continuous service were short,"[39] and this tendency was especially pronounced among female workers.

(3) Working Conditions

Working hours in most factories were usually "from sunrise to sunset," but in the winter, when the demand for matches was greatest, "work continued uninterrupted until about ten o'clock." On average, therefore, the working day was 14 hours long. Elementary-school pupils "came to work early in the morning and worked for one or two hours before going to school, and returned after school to work until sundown." Apparently, therefore, workers laboured only part of the day in exceptional cases. Workers usually had two days off a month.

Wages were paid on a daily basis in the case of men and on a piecework basis in the case of women. This was related to the nature of the work done: lining up the matchsticks, filling boxes, pasting labels, and wrapping (jobs done mostly by women) were paid on a piecework basis, and applying the match composition to the tips, drying the matches, and mixing the composition (done mostly by men) were paid as daily wages.[40] Table 3.11 shows daily earnings at nine factories whose data are reported in *Matchi shokkō jijō*, divided into 10-sen brackets. Match-factory workers were clearly badly paid, especially women and children. In addition to low wages, moreover, working in a match factory involved considerable danger. Fires occurred frequently and each time left many dead and injured, but "not a single factory gave relief to the victims."[41]

Table 3.11. Wages of Workers in Nine Match Factories, Breakdown by Age and Sex, 1901

Daily wage (sen)	Number of workers			Percentage of total	
	Male	Female	Total	Male	Female
<10	110	492	602	9.1	14.7
10–14	157	1,058	1,215	13.0	31.6
15–19	310	1,212	1,522	25.7	36.2
20–29	325	485	810	26.9	14.5
30–39	118	101	219	9.8	3.0
40–49	117	—	117	9.7	—
50–59	55	—	55	4.5	—
60–79	16	—	16	1.3	—
Total	1,208	3,348	4,556	100.0	100.0

Source: *Matchi shokkō jijō* [Conditions of Match-factory Workers], p. 143.

The match-making industry relied on poorly paid labour by young women and children of the urban lower class under terrible working conditions. From the viewpoint of the lower strata of urban society, match-factory work turned poor young women and children into an urbanized working class, giving them a role in Japan's capitalist growth process and changing the labour structure from within without dislocating them from their homes in the slums of Japan's cities.

III. The Post-First World War Period

1. The Urban Lower Class after the Russo-Japanese War

Japanese capitalism came to the end of its industrial revolution and began the transition to monopoly capitalism after the Russo-Japanese War of 1904–1905. It was during this period that the Ministry of Home Affairs (Naimushō, Chihōkyoku) conducted the first serious studies of the urban lower classes. The first, entitled *Saimin chōsa tōkeihyō* [Statistical Tables from a Study of the Poor], was conducted in 1911; the report was released the following year. The second was conducted in 1912 and published in 1914 as *Saimin chōsa tōkeihyō tekiyō* [Synopsis of Statistical Tables from a Study of the Poor]. After the Russo-Japanese War, an arson incident at Hibiya in 1905 sparked a nationwide rash of riots by city dwellers, large numbers of them poor.[42] The Ministry of Home Affairs' studies were part of a government effort to determine the condition of the poor in the cities and to check discontent before it developed into an organized opposition movement. Using these studies, let us first look at the realities of female labour in the urban lower class around 1910. We will then look at the changes that took place after the end of the First World War, using data on Tokyo to compare the two periods.

The populations of Japan's largest cities climbed sharply after the industrial revolution, mainly due to the sharp increase in the number of people employed by industry. Tokyo's population was just below one million until the mid-1880s; by 1905–1910, it had effectively doubled. Half of its inhabitants had come from other parts of Japan, and their official domiciles (*honseki*) were not in Tokyo.[43]

Though the population came to be concentrated in large cities in the course of industrialization, new arrivals did not necessarily begin immediately to work in factories. The majority settled first in urban slums, swelling the ranks of the poor. The exact number of the poor cannot be accurately determined, but data on those officially termed indigent (*saimin*) in 11 wards of Tokyo are shown in table 3.12. The definition of indigent according to the Tokyo city authorities who compiled these figures is: "coolies, cart-drivers, day labourers, etc., who are unable to pay their ward charges (*kuhi*) and earn 20 yen a month or less or pay house rent not exceeding 3 yen."[44] This category of the poor accounted for 12 per cent of the population.

Table 3.12. Distribution of the "Indigent" in Tokyo, 1911

Ward population	Approximate number	Percentage of total, 1908
Fukagawa	30,213	25.4
Honjo	35,000	21.4
Asakusa	69,869	37.6
Shitaya	36,073	28.8
Shiba	3,731	2.7
Azabu	2,622	4.0
Akasaka	500	1.0
Yotsuya	5,458	13.1
Ushigome	1,200	1.3
Koishikawa	18,672	19.8
Hongo	1,398	1.5

Source: Nakagawa Kiyoshi, "Senzen ni okeru toshi kasō no tenkai" [The Evolution of the Urban Lower Class in Pre-war Japan], part 1, *Mita gakkai zasshi*, vol. 7, no. 3.

The indigent were concentrated primarily in wards of Tokyo where, from the 1890s to the first decade of the 1900s, new factories had sprung up and the number of industrial workers was rapidly growing—Honjo, Fukagawa, and Koishikawa—and in the traditional slum areas of Tokyo—Asakusa and Shitaya. These data are therefore in agreement with observations regarding table 3.3 above. Because the increase in the population of the lower class occurred in the industrial zones, it seems justified to interpret the change in the nature of urban lower-class society as having been caused by industrial development. Indeed, the composition of the urban lower class during the period from the Russo-Japanese War to the First World War underwent a change, the most apparent feature of which was the growth in female labour.

Table 3.13 breaks down the occupations of the indigent in four wards of Tokyo on the basis of a 1911–1912 Ministry of Home Affairs survey. Because the wards covered (Shitaya, Asakusa, Honjo, and Fukawa) represent both old and new slum districts, the data give a fair idea of the jobs held by the poor during this period. The preponderance of factory workers is immediately apparent. From 1911 to 1912, the largest category of the poor with jobs, whether heads of households or dependents, male or female, was of those employed in industry. There is no doubt that, compared to the previous period, the predominant occupation of the lower class was now the industrialized type. Women particularly, whether heads of household (mostly widows) or members of male-headed households, were overwhelmingly employed as factory workers.

It is interesting to note that among workers from male-headed households, the number of women is over four times as great as the number of male workers (excluding household heads). The reason is that in almost all

Table 3.13. Occupations of the "Indigent" in Four Wards of Tokyo (Shitaya, Asakusa, Honjo, and Fukagawa) Based on 1911–1912 Survey Data (percentages)

Occupation	Head of household	Female head of household	Employed male dependent	Employed female dependent
Agriculture, fishing	1.1	0	0	0
Manufacturing	34.9	46.8	61.1	82.7
Metals	5.3	0.9	12.1	1.0
Spinning	3.4	11.0	4.3	33.4
Apparel, accessories	6.0	19.0	6.6	24.5
Wooden, bamboo products	5.2	7.0	6.3	2.5
Civil engineering, construction	8.9	0	6.2	1.0
Commerce	14.0	18.7	6.0	6.2
Used, discarded goods		0.6		
Street stalls	4.0	1.5	1.3	1.3
Transportation	22.1	0	6.2	0.3
Rickshaw-pullers	13.1	0	2.6	—
Porters, delivery personnel	6.8	0	1.6	0.1
Civil service, independent	3.2	0.6	5.5	1.3
Housework	0.1	0.3	0.4	0.5
Other occupations	13.8	21.6	13.5	7.7
Day labour	8.6	3.2	4.2	1.5
Refuse-sorting, disposal	2.4	4.7	1.8	1.9
Unemployed	1.9	12.0	1.8	0.1
Total	100.0	100.0	100.0	100.0
Actual number	5,935	342	954	4,005

Sources: Ministry of Home Affairs, Chihōkyoku, *Saimin chōsa tōkeihyō* [Statistical Tables from a Study of the Poor] and *Saimin chōsa tōkeihyō tekiyō* [Synopsis of Statistical Tables from a Study of the Poor]. Cf. Nakagawa Kiyoshi, "Senzen ni okeru toshi kasō no tenkai" [The Evolution of the Urban Lower Class in Pre-war Japan] *Mita gakkai zasshi*, vol. 7, no. 3.

cases they were the householder's spouse. In this survey, 81.3 per cent of employed women (excluding female household heads) were spouses; 84.5 per cent were at least 20 years of age.[45] This speaks for the role of the family and the formation of households in the urban lower class. But for the poor of this period, the "only way to support a small family (the spouses) was to expel from the family at a relatively early age a considerable number of dependents."[46] Minors working and living outside (*higenjūmin*, literally non-resident family members) were almost exclusively the children of such families. Upon completion of elementary school at about 12 years of age, children were sent off on their own and began to live outside the family.

Though urban poor dwellers began forming and maintaining families, this was made possible by expelling an average of 0.40 "non-resident family members" per household.[47]

Whether heads of household or not, female job holders were most often employed in spinning or the manufacture of apparel and accessories. The latter occupation was generally engaged in at home, whereas most workers in the spinning industry commuted to factories in the spinning centres of Honjo and Fukagawa. Tokyo spinning-mill owners had just begun to build workers' dormitories and recruit female workers from remote areas, but the urban lower class continued to furnish a certain amount of mill labour.

The primarily industrial employment structure of the poor of this period was not peculiar to Tokyo; it also applied to Osaka. Table 3.14 shows occupations, monthly wages, and days worked by poor female workers in two wards of Tokyo and four districts (*machi*) of Osaka. In Osaka, industry was not centred on textiles as in Tokyo; the majority of female workers were employed in the chemical industry (mainly match making) and apparel and accessory manufacture. In Osaka, as in Tokyo, therefore, the main branch of female employment was industrial.

Wages and days worked by women are also shown in table 3.14. In both Tokyo and Osaka, factory workers worked about the same number of days per month or slightly more than in other occupations, but wages (monthly) were lower. Though detailed data on daily working hours are unavailable and a simple comparison cannot be made, it is known that 71.6 per cent of the wives of the indigent were employed,[48] apparently because most women of the lower class had to have extra income to make ends meet. Only if their low wages were added to their husbands' could the urban poor manage to eke out an existence.

Finally, we should address the status of the "non-resident family members" (*hi-genju jinkō*), the majority of whom were employed minors between 10 and 19 years of age. By far the majority of girls in this category were employed as maids or shop-helpers (*tetsudai*); boys worked mainly as factory workers in the metal and machinery industries. This implies that though they were still under severe constraints, men were on their way towards escaping from their class conditions by working for a big company in heavy industry.[49]

2. Metamorphosis of the Urban Lower Class: The Post-First World War Period

Government studies on the conditions of the poor, begun in 1911–1912, were launched again a decade later, after the First World War.[50] Again, it was popular unrest, culminating in an outbreak of riots known as the rice riots of 1918, which spurred public authorities to action in 1920–1921. But though the events that triggered the poverty surveys of 1910–1911 and 1920–1921 were similar, their findings diverged quite substantially.

In Tokyo's case, the "indigent" had accounted for 12.6 per cent of the

Table 3.14. Occupations of Female Job-holders, 1912

	Number				Average monthly wage (yen)				Average days/month			
	Tokyo[a]		Osaka		Tokyo[a]		Osaka		Tokyo[a]		Osaka	
	HH[b]	ED[b]	HH	ED	HH	ED	HH	ED	HH	ED	HH	ED
Fishing	0	2	0	0	—	3.10	—	—	—	15.0	—	—
Ceramics	4	35	6	46	6.59	4.55	6.12	3.44	26.2	25.1	28.4	24.9
Metal industry	3	32	1	14	6.64	4.23	4.80	3.42	24.3	24.8	24.0	25.6
Machinery, instruments	0	6	0	4	—	4.57	—	4.09	—	27.0	—	24.8
Chemicals	3	53	12	219	3.18	3.87	4.16	3.07	24.3	24.6	26.1	24.7
Disposal of animal carcases	0	9	10	32	—	3.68	4.94	2.39	—	24.1	25.3	23.7
Textile industry	27	998	12	97	3.09	1.79	3.24	3.80	26.7	25.2	26.1	24.0
Dyeing	2	13	0	1	6.50	4.45	—	3.00	20.0	23.1	—	25.0
Paper, rubber	0	147	3	69	—	1.72	4.07	3.68	—	25.1	—	26.4
Wood, bamboo	3	34	12	67	6.55	3.80	3.11	2.61	28.3	23.2	28.3	25.1
Comestibles	1	31	2	100	4.00	4.92	4.33	4.23	22.0	25.9	26.3	24.5
Apparel, accessories	24	457	82	291	4.88	2.41	3.36	2.39	2.72	24.1	25.5	24.2
Building	0	39	0	2	—	4.21	—	6.25	—	18.1	—	24.5
Printing	1	10	0	1	13.00	5.83	—	3.75	20.0	25.4	—	25.0
Toys, etc.	2	57	1	55	1.65	2.34	8.00	2.58	15.0	24.6	28.0	24.4
Pipe-making, etc.	0	4	1	2	3.00	5.00	4.20	2.77	30.0	25.0	28.0	29.0
Commodity sales	19	79	23	49	7.10	4.49	5.35	4.43	27.6	24.9	25.4	27.6
Used, discarded goods	2	1	19	64	10.75	10.75	6.52	5.25	25.0	—	25.3	22.0
Hair-styling	3	9	2	0	4.50	3.99	5.75	5.75	25.0	21.2	27.7	—

Merchant marine	0	1	0	2	—	6.00	—	—	—	20.0	—	25.0
Porter	0	2	0	5	—	1.73	—	2.00	—	15.0	—	20.0
Regular manual labour	2	20	3	4	5.77	5.27	3.57	3.08	24.5	23.6	23.3	21.3
Massage	1	4	14	7	6.00	3.70	3.88	3.48	23.0	21.3	26.4	22.7
Entertainment	1	4	0	8	5.00	4.90	—	5.20	20.0	27.5	—	22.8
Cleaning, weeding	0	7	1	2	—	3.37	3.00	2.27	—	24.3	30.0	12.5
Refuse-gathering and disposal	2	10	9	14	6.22	3.58	6.29	5.51	26.5	24.5	24.9	24.6
Day labour	6	38	4	12	7.20	4.16	3.00	2.06	22.5	18.9	14.3	20.9
Total (including other categories)	126	2,172	261	1,294	5.24	2.63	4.24	3.25	25.9	24.5	25.6	24.5

a. Honjo and Fukagawa wards.
b. HH = household head; ED = employed dependent.
Source: Ministry of Home Affairs, Regional Affairs Bureau, *Saimin chōsa tōkeihyō tekiyō* [Synopsis of Statistical Tables from a Study of the Poor].

population in 1911–1912, but by the 1920 survey they had dwindled to a mere 3.4 per cent.[51] It was not because the criteria for indigence had become any more severe: indeed, they were the same, based mainly on income, during both periods.[52] It was simply that far fewer people were under the poverty line.

From about 1917 (midway through the First World War) until March 1920, when the first post-war panic halted the wartime economic boom, the Japanese economy prospered and industrial production soared. Enterprises fought to get the workers they needed to increase production. There was a generalized labour shortage, to the advantage of workers, whose numbers grew remarkably, especially in the metal and machinery industries, where they grew 1.7-fold between 1914 and 1919.[53] It was mentioned above that many poor male non-resident family members were employed in the metal and machinery industries in 1911–1912; throughout the First World War boom, a certain number of the urban poor, notably these poor male non-resident family members, managed to rise from the lower class. The result was both quantitative shrinkage and major qualitative transformation of urban poor society. The changes are described below.

First, poverty became geographically decentralized. In Tokyo, its centre of gravity shifted from the traditional poor quarters of late feudal times—Asakusa and Shitaya—to Fukagawa and Honjo. And despite the prevalent indigence of these four wards, their share of Tokyo's total indigent population declined. Thus the number of poor shrank at the same time that poor people began to settle in different districts of Tokyo. "There was gradual improvement in the poor quarters in the centre of the city, and the poor began to move from the centre to the outskirts, then from the outskirts to the suburban and rural districts outside the city."[54]

Second, workers in big factories, particularly in heavy industry, rose from the ranks of the poor. Eventually a clear line was drawn between workers in heavy industry and the lower strata of urban society according to housing, income level, and participation of family members in several other occupations.[55] Those factory workers who were still seen in the ranks of the urban lower class were henceforth employed by small firms.

Third, the urban lower class stopped passing on its poverty to the younger generation. During this period, only 8 per cent of the indigent had had indigent parents.[56] Industrial workers, especially workers in big factories, born in urban poverty of many generations' standing, were breaking away from their class origins, with the result that "the younger generation was no longer locked into the self-perpetuating cycle of poverty breeding poverty in the cities."[57]

Fourth, housing improved. Cheap lodging-houses, where the bulk of urban poor families had lived until the first decade of the 1900s, became primarily the temporary lodgings of unmarried young men, and ceased to house families. Families moved to row houses, which evolved from their original form of communal tenements to apartment houses with a modicum of privacy.

Table 3.15. Occupations of the Indigent, 1920–1921 (percentages)

Category Subcategory	Head of household (10 wards[a])	Head of household (3 wards[b])	Spouse (3 wards)	Employed dependent other than spouse	
				Male (3 wards)	Female (3 wards)
Agriculture, fisheries	1.5	2.6	—	2.2	—
Mining, industry	35.2	47.9	61.2	74.4	85.3
Ceramics, quarrying	—	2.8	3.0	18.9	8.0
Metals	3.6	4.8	0.5	12.2	1.3
Spinning	—	1.2	10.9	2.2	4.0
Apparel, accessories	4.8	6.4	35.3	2.2	26.7
Engineering/public works	7.4	24.1	3.5	12.2	—
Factory worker	15.7	—	—	—	—
Commerce	7.8	12.1	11.9	8.9	4.0
Used goods, peddling, street stalls	7.8	—	—	—	—
Transportation	18.2	20.1	2.5	6.7	1.3
Rickshaw puller	10.9	—	—	—	—
Porter	6.6	—	—	—	—
Public service, independent	1.2	3.8	3.5	3.3	—
Housework	—	—	1.5	—	6.7
Other industries	36.1	13.5	19.4	4.4	2.7
Day, ordinary labourer	31.3	—	—	—	—
Miscellaneous	3.2	—	—	—	—
Total	100.0	100.0	100.0	100.0	100.0
Actual number	2,395	497	201	90	75

a. Kyobashi, Shiba, Azabu, Akasaka, Yotsuya, Ushigome, Shitaya, Asakusa, Honjo, and Fukagawa.
b. Yotsuya (Asahimachi), Asakusa (Asakusamachi), and Fukagawa (Honmura-machi, Sarue-urachō).
Source: Nakagawa Kiyoshi, "Senzen ni okeru toshi kasō no tenkai," [The Evolution of the Urban Lower Class in Pre-war Japan], part 2, *Mita gakkai zasshi*, vol. 71, no. 4: 105, table 20.

In the period during which these changes took place (1920–1921), the employment structure of the urban lower class also changed (table 3.15). First, the percentage of male heads of household employed in industrial occupations had declined compared to 1911–1912, while that of manual labourers (coolies, rickshaw-drawers, porters, etc.) had increased. The same shift from industrial to manual labour applied to their wives. Though the majority of wives with jobs (about 60 per cent) were still employed in industry, the percentage engaged in transportation and "other industries" employing manual labour rose in comparison with the previous study period, 1911–1912.

In discussing wives' occupations, it must be borne in mind that the percentage of employed wives dropped substantially after the First World War, though this factor is not indicated in table 3.15. In 1911–1912, 70 per cent of wives of indigent men held outside jobs; in 1921, the rate was 44 per cent.[58] To be sure, a greater percentage of urban lower-class wives held jobs than did the wives of workers employed by major heavy industrial companies. But the difference was nevertheless clear with respect to the period when wives had to have their own income to make ends meet. The drop in married women's employment rate is presumed to result from a rise in the income of heads of households. As a result of the wartime labour shortage and a strong labour movement from the middle of the war period on, factory workers' wages rose in both absolute and relative terms from 1918–1919. The incomes of the poor "followed this upward trend" in factory workers' wages, but with a certain time-lag; around 1920–1921 the poor "sought to benefit from wartime changes."[59] Under these circumstances, employed wives' incomes rose with respect to 1911–1912 levels, and the average number of days worked sank to 20 a month.

The percentage of employed detached family members besides wives hardly changed, and, as in 1911–1912, the major form of employment was industrial. More women were employed in the manufacture of apparel and accessories, done mainly at home, than in any other type of work. The number of male factory workers was constant. Non-spouse employed dependents, probably mostly children of the head of household and his spouse, were therefore presumably able to escape from the category of urban lower class.

The last observation concerns family size: in 1921, the average size of an indigent household was 4.1 people,[60] whereas the 1911–1912 figure was 3.5. The reason for this increase appears to be that more children ceased to leave the family to work. Before, it had been too difficult to support children; after the First World War, however, it was easier for the family to remain together.

IV. Conclusion

Following the period we have just considered, a number of changes, addressed briefly below, affected the situation of the urban poor. The first affects the number of poor. Until the mid-1920s, the lower class continued to represent about the same percentage of the population as in the immediate post-First World War period, but then its numbers took an upswing. This recrudescence of poverty was caused by the financial panic of 1927 and the depression (known as the "Shōwa Panic") that followed in late 1929. Bankruptcies and a worsening economic climate affected workers at big factories, proprietors of small independent businesses, and tenant farmers. In 1932–1933, the percentage of the population in the lower class again rose above 10 per cent.[61] Surveys conducted during the Shōwa Panic period defined the lower class mainly in terms of their lack of money for paying living expenses, whereas in previous surveys by government authorities two sorts of criteria had defined poverty: the ability to pay living expenses and the subject's residential environment, namely in substandard group housing in the slums. The latter criterion was abandoned in an apparent response to the geographical spread of such people. Indeed, a survey conducted in Tokyo during the Shōwa crisis indicates that the main pockets of poverty remained in Fukagawa and Honjo wards, but the poor had come to live everywhere else in the city as well.[62] In their new settlements, the lower class lived in row houses and one- to four-dwelling apartments that were somewhat more spacious than in the past.

The occupations of the urban lower class during the Shōwa crisis were nearly evenly divided among four categories: industry, manual labour, miscellaneous employment, and unemployment.[63] This signified a decrease in industrial and manual labour, and a corresponding falling back on odd jobs and "no job at all." In other words, those who lost their jobs in industry and manual labour did not necessarily become totally unemployed; a very large proportion of them performed odd jobs, whose availability reduced the direct impact of lost jobs on unemployment figures. However, even odd-job holders were far from able to pay their living expenses, and the majority were regarded by the government as needing assistance. The greatest number of female heads of households were employed in industry, the next greatest in sales. The breakdown of industrial occupations shows that the vast majority were active in "various industries" such as sewing/tailoring and the manufacture of footwear,[64] and this probably applied equally to the wives of the poor. Women's occupations remained the same as before. During this period, wives and children again strongly desired to find jobs of whatever kind, even at low wages, because the family budget could not be balanced otherwise.

In sum, during the ensuing economic crisis, a serious reversal took place, again swelling the ranks of the poor and aggravating their living conditions, which had undergone a measure of improvement in the wake of the period of rapid economic growth stimulated by the First World War. The situation of urban lower-class working women during the war economy put in place after the Manchurian Incident (1931)—the next phase of the history of the Japanese urban poor—deserves to be addressed elsewhere.[65]

Notes

1. Tsuda Masumi, "Nihon no toshi kasō shakai," [Japan's Urban Lower Class], *Keizaigaku ronshū*, vol. 24, no. 2 (1956); Tsuda Masumi, *Nihon no toshi kasō shakai*, [Japan's Urban Lower Class] (Mineruba Shōbō, Tokyo, 1972).

2. Sumiya Mikio, *Nihon chinrōdō shiron* [Historical Essay on Wage Labour in Japan] (Tokyo University Press, Tokyo, 1955); Sumiya Mikio, *Nihon no rōdo mondai* [Japanese Labour Problems] (Tokyo Daigaku Shuppankai, Tokyo, 1967).
3. Hyōdō Tsutomu, *Nihon ni okeru rōshi kankei no tenkai* [The Evolution of Capital and Labour in Japan] (Tokyo Daigaku Shuppankai, Tokyo, 1971).
4. Nakagawa Kiyoshi, "Senzen ni okeru toshi kasō no tenkai" [The Evolution of the Urban Lower Class in Pre-war Japan], parts 1 and 2, *Mita gakkai zasshi*, vol. 71, nos. 3, 4 (1978); Nakagawa Kiyoshi, "Zoku senzen ni okeru toshi kasō no tenkai" [Sequel to the Evolution of the Urban Lower Class in Pre-war Japan], *Shōgaku ronshū* (Niigata University), 13 (1980); Nakagawa Kiyoshi, "Kantō daijishin chokugo no toshi kasō" [The Urban Lower Class in the Aftermath of the Great Kantō Earthquake], *Shōgaku ronshū*, 15 (1982).
5. "Urban lower class" in this chapter refers to the group in a restricted sense. Readers should refer to Hashimoto Norichika's *Chihō toshi no kasō minshu to minshu bōdō* [The Lower Class of Regional Urban Centres and Popular Uprisings] (United Nations University, Tokyo, 1980) for materials on the lower class in outlying urban areas.
6. Included in *Tokyo-shi shikō*, 48 (1959). My citation, however, is from Ishizuka Hiromichi, *Tōkyō no shakai keizaishi* [The Social Economic History of Tokyo] (Kinokuniya Shoten, Tokyo, 1977).
7. Yokoyama Gennosuke, *Nihon no kasō shakai* [Japan's Lower Classes] (Iwanami Bunko, Tokyo, 1949), p. 22.
8. Ishizuka, *Tokyo no shakai keizaishi*, pp. 23–24.
9. Matsumoto Shirō, "Bakumatsu, ishin-ki ni okeru toshi no kōzō" [Urban Infrastructure during the Late Feudal and Early Meiji Periods], *Mitsui Bunko Ronsō*, 4 (1970).
10. Miyachi Masato, *Nichi-Rō sengo seiji-shi no kenkyū* [The Political History of the Post-Russo-Japanese War Period] (Tokyo Daigaku Shuppankai, Tokyo, 1973), pp. 193–200.
11. Yoshida Kyūichi, "Edo jidai no toshi kasō shakai" [The Urban Underclass in the Edo Period], *Nihon Shakai Jigyō Daigaku kenkyū kiyō*, vol. 27 (1981), p. 29.
12. Ibid.
13. Sumiya, *Nihon chinrōdō shiron*, pp. 83–89.
14. Ibid.
15. Unno Fukuju, "Genchikuron" [Theory of Primitive Accumulation], in Ishii Kanji, Unno Fukuju, and Nakamura Masanori, eds., *Kindai Nihon keizaishi wo manabu* [Learning from Modern Japanese Economic History], vol. 1 (Yūhikaku, Tokyo, 1977), p. 17.
16. Daiga Koji, *Hintenchi kikankutsu tankenki* [Exploring the Cold and Starving Slums], 1893; reprinted in Nishida Taketoshi, ed., *Meiji zenki no toshi kasō shakai* [The Urban Lower Class in the Early Meiji Period] (Kōseikan, Tokyo, 1970), 108.
17. Sumiya, *Nihon chinrōdō shiron*, p. 100.
18. Ishizuka Hiromichi, *Toshi kasō shakai to saimin jukyōron* [The Urban Lower Class and the Living Status of the Indigent] (United Nations University, Tokyo, 1979).
19. Rōdō Undo Shiryō Iinkai [Labour Movement Reference Committee], ed., *Nihon rōdō undō shiryō* [Historical References to the Japanese Labour Movement], vol. 1; reprinted in 1962.

20. *Kokumin shimbun*, 29 October and 5 November 1893; reprinted in *Nihon rōdō undō shiryō*, vol. 1.
21. Reprinted in Nishida, *Meiji zenki no toshi kasō shakai*.
22. At the 12 December 1981 research meeting of the United Nations University project team, Dr Helena Sumiko Hirata of Brazil pointed out that the collection of trash had become a means of earning income from early in the modern period. Although it would be extremely interesting to study the extent to which trash recyling technology developed in Japan, this question will not be taken up in this study.
23. *Tokyo no hinmin* serial in *Jiji shinpō*, 1896; reprinted in Hayashi Hideo, ed., *Ryūmin* [Migrants], vol. 4 (Shinjinbutsu Oraisha, 1971).
24. Oishi Kaichirō, "Kadai to hōhō" [Tasks and Methods], in Oishi Kaichirō, ed., *Nihon sangyō kakumei no kenkyū* [Studies on the Japanese Industrial Revolution], vol. 1 (Tokyo Daigaku Shuppankai, Tokyo, 1975).
25. Ishizuka Hiromichi, *Nihon shihon shugi seiritsushi kenkyū* [Studies on the History of the Establishment of Japanese Capitalism] (Yoshikawa Kōbunkan, Tokyo, 1973), pp. 315–330.
26. Sumiya Mikio, *Nihon no rōdō mondai* [Japan's Labour Problem] (Tokyo Daigaku Shuppankai, Tokyo, 1967), pp. 66–72.
27. At this point in the chapter, the manner of presentation changes, because data are not available on the overall conditions of the urban lower class in specific regions or in the country as a whole. It is only extrapolation of the information in part I and section 1 of part III that allows us to reconstruct the total numbers and the geographical location, occupational composition, and family conditions of the urban poor. Studies of the match industry do give us, however, a specific but partial picture of conditions in urban lower-class society. Originally, a study was also made of the spinning industry, but space does not permit its inclusion here. The most comprehensive treatment thus far of female labour in the spinning industry during the period under consideration is that of Murakami Hatsu, "Sangyō kakumeiki no joshi rōdō" [Female Labour in the Industrial Revolution], in Joseishi Sōgō Kenkyūkai [History of Women Research Group], ed., *Nihon joseishi* [The History of Japanese Women], vol. 4: *Gendai* [The Modern Era] (Tokyo Daigaku Shuppankai, Tokyo, 1982).
28. Takamura Naosuke, "Sangyō-bōeki kōzō" [The Industry and Trade Structure], in Ōishi, *Nihon sangyō kakumei no kenkyū*.
29. Komiyayama Takuji, *Nihon chōshō kōgyō no kenkyū* [Studies on Japanese Small Industrial Enterprises] (Chūō Kōronsha, Tokyo, 1941), p. 142.
30. Yokoyama, *Nihon no kasō shakai*, p. 129; Agriculture and Commerce Ministry, *Zenkoku kōjō tōkeihyō* [Nationwide Factory Statistics Table] (annual).
31. Yamashita Naoto, "Keiseiki Nihon shihon shugi ni okeru matchi kōgyō to Mitsui Bussan" [The Match Industry and Mitsui Bussan during the Formative Period of Capitalism in Japan], pp. 100–105; also the source for the following material.
32. Ministry of Agriculture and Commerce, Commerce and Industry Bureau, ed., *Shokkō jijō* [Conditions of Workers] (1903; reprinted in 1967), p. 129; hereafter this report is referred to as *Matchi shokkō jijō* [Conditions of Match Factory Workers].
33. Ibid.
34. Ibid.
35. Rōdō Undō Shiryō Iinkai [Labour Movement Reference Committee], *Nihon rōdō undō shiryō* [Historical References to Japanese Labour], vol. 10 (Rōdō

Undō Shiryō Kankō Iinkai, Tokyo, 1959), p. 270.
36. *Matchi shokkō jijō*, p. 167.
37. Ibid., p. 136.
38. Ibid., pp. 136–138.
39. Ibid.
40. Ibid., pp. 134–139.
41. Ibid., p. 151.
42. Miyachi Masato, "Nichi-Rō zengo no shakai to minshū" [Society and the People before and after the Russo-Japanese War], in Rekishigaku Kenkyūkai and Nihonshi Kenkyūkai, eds., *Kōza Nihonshi* [Lectures on Japanese History], vol. 6: *Nihon teikoku shugi no keisei* [The Emergence of Japanese Imperialism] (Tokyo Daigaku Shuppankai, Tokyo, 1970).
43. Ishizuka, *Tōkyō no shakai keizaishi*, pp. 101–102.
44. Nihon Shakai Jigyō Daigaku Kyūhin Seido Kenkyūkai [Japan College of Social Work Study Group on the Poverty Relief System], ed. *Nihon no kyūhin seido* [Japan's Poverty Relief System] (Keisō Shobō, Tokyo, 1960), p. 153.
45. Ministry of Home Affairs, Regional Affairs Bureau, *Saimin chōsa tōkeihyō* [Statistical Tables from a Study of the Poor], pp. 21–24; *Saimin chōsa tōkeihyō tekiyō* [Synopsis of Statistical Tables from a Study of the Poor], pp. 16–17.
46. Nakagawa, "Senzen ni okeru toshi kasō no tenkai," part 1.
47. Ibid.
48. Ministry of Home Affairs, Regional Affairs Bureau, *Saimin chōsa tōkeihyō tekiyō*, pp. 16–17.
49. Hyōdō, *Nihon ni okeru rōshi kankei no tenkai*, pp. 315–20.
50. In Tokyo's case, such studies included Tōkyō-shi Shakaikyoku [Tokyo Metropolitan Social Affairs Bureau], *Tōkyō shinai no saimin ni kansuru chōsa* [A Survey of the Poor in Tokyo] (survey conducted in 1920, report published in 1921); and Ministry of Home Affairs, Shakaikyoku, *Saimin chōsa tōkeihyō* [Statistical Tables from a Study of the Poor] (survey conducted in 1921, report published in 1922).
51. Tokyo-shi Shakaikyoku, *Tōkyō shinai no saimin ni kansuru chōsa*, pp. 4–5.
52. Nakagawa, "Senzen ni okeru toshi kasō no tenkai," part 2, pp. 95–96.
53. Ninomura Kazuo, "Rōdōsha kaikyū no jōtai to rōdō undō," *Iwanami kōza Nihonshi*, vol. 18 (Iwanami Shoten, Tokyo, 1975), p. 105.
54. Tokyo-shi Shakaikyoku, *Tōkyō shinai no saimin ni kansuru chōsa*, p. 23.
55. Hyōdō, *Nihon ni okeru rōshi kankei no tenkai*, pp. 442–79. Big heavy industrial plants were then at a turning point, and beginning to hire part-time workers and have work done by outside subcontractors. What drew the line between workers in these factories and the urban lower class was the permanence of workers' status in big factories.
56. Tokyo-shi Shakaikyoku, *Tōkyō shinai no saimin ni kansuru chōsa*, pp. 93–94.
57. Nakagawa, "Senzen ni okeru toshi kasō no tenkai," part 2, p. 99.
58. Ibid., p. 108.
59. Ibid.
60. Tokyo-shi Shakaikyoku, *Tōkyō shinai no saimin ni kansuru chōsa*, pp. 23–24.
61. Nakagawa, "Zoku senzen ni okeru toshi kasō no tenkai," p. 29.
62. Tokyo-shi Shakaikyoku, *Tōkyō shinai yōkyūgosha ni kansuru chōsa* [A Survey of People Requiring Relief in Tokyo] (survey conducted in November 1931, published in February 1932), fig. 1, etc.
63. Nakagawa, "Zoku senzen ni okeru toshi kasō no tenkai," p. 31.

64. Tokyo-shi Shakaikyoku, *Tōkyō shinai yōkyūgosha ni kansuru chōsa*, pp. 42–65.
65. This chapter limits consideration to the ethnic Japanese urban lower class, but, beginning in the middle of the First World War, the number of Koreans and other colonized nationalities in the lower strata of Japanese urban society also increased rapidly. To date, the most detailed and systematic treatment of the problems of this segment of the urban poor is found in two works, Matsumura Takao's "Nihon teikoku shugi ka ni okeru shokuminchi rōdōsha" [Labourers from the Colonies under Japanese Imperialism], in Keio University's *Keizaigaku nenpō*, vol. 10 (1966) and Totsuka Hideo's "Nihon ni okeru gaikokujin rōdōsha mondai ni tsuite" [The Foreign Worker Problem in Japan], in the University of Tokyo's *Shakai kagaku kenkyū*, vol. 25, no. 5 (1974).

Chapter———4

Family-run Enterprises: An Overview of Agriculture and Fisheries

Kazutoshi Kase

After the difficult years of reconstruction following the defeat in the Second World War, Japanese capitalism entered a period of rapid economic growth beginning in about 1955. The economy, centred on the heavy and chemical industries, expanded and productivity increased dramatically. Japan emerged from the desolation and deprivation of a defeated nation to become one of the world's leading industrial powers. This rapid development stimulated brisk growth in the numbers of workers employed in capital-based companies, and paralleling this growth, the number of employees in family enterprises dropped sharply, and this sector lost its predominant position among the working population. The large-scale enterprises pushed aside family-run businesses and made it difficult for them to secure labour.

These historical factors provide the background to the discussion in this chapter. The introduction of technological changes in family enterprises occurred in response to the depletion and ageing of labour resources under conditions of economic growth. Agriculture and fisheries, the most important types of family enterprise in Japan, have strong market limitations. In these areas, new technologies are linked not so much to increased output as to filling the gaps left by fewer and older workers.

Changes in employment conditions vary greatly for men and women in family-run businesses. In general, women (particularly those who are married) are bound to domestic and child-raising duties. This means that the trend toward work in family-run enterprise is stronger among women than among men. Whether or not that trend manifests itself or not depends on the particular conditions of owner-managed businesses, and here agriculture and fisheries show contrasting trends. To oversimplify the matter drastically, in agriculture there tends to be an exodus of men, with women left to assume responsibility for the majority of the work, while in fishing men remain indispensable for the work on the boats, forestalling a similar pattern. Comparisons of this kind are helpful in this study of the response of family enterprises to the expansion of the labour market and technological change.

Because agriculture exhibits relatively straightforward patterns of change and has been the subject of many previous surveys and studies, this sector will be touched on only briefly; the bulk of the chapter describes owner-managed fishery concerns.

I. Changes in the the Agricultural and Fisheries Industries and Patterns of Employment[1]

Two new features defined conditions in the agricultural and fisheries industries during the early post-war reconstruction period. First, as a result of demobilization and the destruction of the economic infrastructure, the former inhabitants of farming and fishing villages returned in huge numbers, creating a massive oversupply of labour. Returnees included kin who had left rural areas to work in the cities or in occupied territories overseas or to enter the military forces. Most took up farming or fishing while they waited for employment opportunities to revive. Second, (Occupation-led) agricultural land reform and reform of fisheries were undertaken in order to enable families to achieve economic independence through their own labour. These trends coincided with the promulgation of Japan's new Constitution and the setting up of labour laws. The net result was a shift away from hierarchical and communal restrictions and the development of conditions that allowed for individual management decisions.

Subsequent rapid expansion in the labour market led naturally to a flow of labour resources away from agriculture and fisheries, as people compared the income from family enterprises with the higher levels possible in other jobs. The first to leave the agricultural and fisheries labour force were the male kin who had represented excess labour. The period of rapid economic growth that began in earnest around 1960 attracted even sons who were in direct line to inherit family businesses, and this outflow occurred in both farming and fishing villages. This demographic trend differed only in specifics: whether workers continued to live in the village and commute to their urban jobs or whether they moved completely out of the villages; whether they were new university graduates, mature adults, or older people; and whether they were male or female.

The impact of this exodus of labour varied considerably between the agricultural and fisheries industries. (Differences can even be detected in different agricultural and fishery activities.) As the number of agricultural workers dropped sharply and the number of farming families decreased, it became increasingly common for workers to hold full-time jobs as well as manage the family business (table 4.1).

A small number of full-time farming households are large-scale agricultural concerns or are made up of elderly persons with no opportunities for other employment. In fisheries, small operations in which fishing is a secondary activity tend to be abandoned, and there is a parallel move away from farming, once the main source of secondary income for fishermen. This has

Table 4.1. Farming Families and Fluctuations in Income Composition, 1955–1980

	1955	1960	1965	1970	1975	1980
Total no. of farming families (thousands)	6,043	6,057	5,665	5,402	4,953	4,661
Composition by income source						
Full-time[a]	34.9	34.3	21.5	15.6	12.4	13.4
Type 1[b]	37.6	33.6	36.7	33.6	25.4	21.5
Type 2[c]	27.5	32.1	41.7	50.8	62.1	65.1

a. "Full-time" indicates farming households with no members who work part-time or full-time in other jobs.

b. "Type 1" indicates households in which income from the family-owned or managed farm is higher than that from other jobs.

c. "Type 2" indicates households in which income from family-owned or managed farm is less than that from other jobs.

Source: Statistics and Information Department, Ministry of Agriculture, Forestry, and Fisheries, *Nōgyō sensasu* [Agricultural Census].

Table 4.2. Owner-managed Fishing Businesses and Composition by Income Source, 1953–1978[a]

	1953[b]	1963	1968	1973	1978
Total (households)	235,761	262,518	248,323	224,968	210,123
Composition by income source					
Full-time	14.3	15.9	20.2	21.6	25.5
Type 1	43.0	51.5	45.5	45.5	42.8
Type 2	42.7	32.7	34.4	32.9	31.8

a. "Owner-managed fishing businesses" refers to households engaged in fishing for the purpose of sale and spending a minimum of 30 days per year at sea.

b. The total for 1953 does not include owner-managed businesses that did not use fishing vessels. The higher totals for other years include such businesses.

Source: Statistics and Information Department, Ministry of Agriculture, Forestry, and Fisheries, *Gyogyō sensasu* [Fisheries Census].

strengthened the position of families engaged solely in fishing. The result is that roughly 70 per cent of fishing families gain their income either solely from fishing or *mainly* from fishing (table 4.2).

The economic situation of households at each stratum by scale of business corresponds to this spread of income sources. The share of farm-generated income in farming families drops noticeably in each stratum. The families in the lowest stratum, those operating the 70 per cent of farms throughout Japan (excluding Hokkaido) that are under one hectare in area, depend on

Table 4.3. Degree of Dependence on Farm Income (farm families, by size of operation, 1950–1980) (percentages)

Size	1950	1960	1970	1980	Percentage of farm families
Less than 0.5 ha	40.5	21.4	9.3	3.3	42.3
0.5–	66.9	52.0	29.0	13.4	28.7
1.0–	78.5	72.0	51.6	29.4	14.4
1.5–	85.5	79.4	65.8	42.2	7.2
2.0–	89.0	87.3	77.4	61.1	7.4
Average for all strata	67.5	52.2	35.0	20.0	100.0

Source: Dependence on farm income is extracted from Statistics and Information Department, Ministry of Agriculture, Forestry, and Fisheries, *Nōka keizai chōsa hōkoku* [Report on Surveys of the Economic Situation in Farming Families]. Breakdown of families based on Statistics and Information Department, Ministry of Agriculture, Forestry, and Fisheries, *Nōgyō sensasu* [Agricultural Census].

farming for only 10 to 20 per cent of their income (table 4.3). By contrast, fishing families of the lowest stratum, those using power-engine driven boats of less than one ton, derive more than 40 per cent of their income from fishing (table 4.4).

The contrasts discussed above are a result of the different ways family-run businesses in these two sectors perform under capitalism. The vulnerability of family-run agricultural operations is immediately apparent, while in fisheries the peculiar characteristics of the industry tend to suppress the emergence of weaknesses. Here, let us look at the conditions that sustain the predominance of single-occupation fishing operations among coastal families.

First are the peculiarities of the labour involved in fishing. While it is relatively easy for farmers to incorporate agricultural tasks into gaps in the schedule of non-farm work, fishermen labour under time restraints imposed by external factors, namely the behaviour of their catch. Fish that are attracted to light, for example, have to be caught at night. The physical demands of long periods of continuous work at sea also limit labour patterns in the fisheries sector. With current levels of technology, only men are able to perform the tasks required of fishing (although women often work in auxiliary roles on fishing vessels as part of husband-and-wife teams.). Owner-managed coastal fishing operations must therefore employ at least one male, and this requirement effectively prevents the head of a fishing household from also taking a non-fishing job,[2] leading to the preponderance of single-occupation specialization in this sector.

Second, as reflected in the fact that the main industry on remote islands is fishing, the development of the labour market in coastal fishing villages is

Table 4.4. Degree of Dependence of Fishing Income (fishing families, by size of fishing vessel, 1951–1980) (percentages)

	1951	1962	1970	1980
Non-power-driven vessels	50.4	34.9	21.7	—
Under 3 tons[a]	71.6	62.6	66.5	
Under 1 ton	—	—	50.0	41.2
1–3 tons	—	—	67.0	49.1
3–5 tons	76.6	62.8	79.4	64.9
5–10 tons	84.8	55.0	87.1	69.6
Laver (*nori*) cultivation	52.4	57.2	68.1	54.3

a. Tonnage figures used for ranking refer to total tonnage of vessels used by each business.

Source: Statistics and Information Department, Ministry of Agriculture, Forestry, and Fisheries, *Gyogyō keizai chōsa hōkoku (Gyoka no bu)* [Survey Report on the Fisheries Economy (Section on Fishing Households)].

comparatively much slower than for the average agricultural village. This slow development means there are few opportunities for wage-based employment in these areas, so that those who wish to become wage-earners must leave the villages. Young people responsible for carrying on a family business, however, cannot leave.

Third, although technological development has occurred in navigation and fishing, what advances have been made have not led, as in agriculture, to increased productivity. Unchanging productivity has meant higher unit prices. Undeniably, the total catch by the Japanese fisheries industry has risen from about 4 million tons before the Second World War (during the 1930s) to 10 million tons in the years since 1970. The source of this growth, however, is mostly catch (destined as animal feed and processed fish products) taken by larger, capital-based enterprises operating in distant and offshore fishing grounds. The size of catches by coastal fishing families has remained constant at around 2 million tons. The high component of high-value, fresh fish in this coastal catch, combined with the rising purchasing power of Japanese consumers resulting from economic growth, has secured a sound income for coastal fishing families as prices for their produce have risen.

Thus, while seeking secondary sources of income has not been a feasible option for fishing households, they have enjoyed economic conditions that have made it possible to earn a satisfactory living from fishing alone. Consequently the joint pursuit of fishery with wage-based jobs has been much slower to develop among people engaged in the fishing industry than among those in agriculture.

Table 4.5. Employment of Farming Household Members, 1960–1980

	Family members over 16 yrs (thousands)	Components (%)			
		Members working on family farm			
		Total	(Share working principally farm jobs)[a]	Members in non-farm jobs only	Members not working
Men					
1960	10,694	79.5	51.8	10.9	9.6
1965	9,816	76.3	42.6	12.1	11.6
1970	9,347	81.3	34.5	9.2	9.5
1975	8,575	79.4	25.4	9.4	11.2
1980	8,228	78.1	24.7	10.5	11.4
Women					
1960	11,683	78.1	52.3	5.2	16.7
1965	10,783	73.8	44.1	7.1	19.1
1970	10,267	76.7	37.3	6.7	16.6
1975	9,350	72.7	27.5	7.3	20.0
1980	8,859	69.0	23.6	9.1	21.9

a. "Share principally working in farm jobs" refers to family members who usually work and for whom the principal content of their job is farm tasks.
Source: *Nōgyō sensasu* [Agricultural Census].

Let us now consider the employment situation in farming and fishing families. As indicated in table 4.5, the percentage of male and female family members engaged principally in farm tasks fell, reaching almost 20 per cent in 1980. None the less, men involved to some extent, however small, in farm tasks account for 80 per cent of the total, and women 70 per cent. The technological advances in agriculture have not severed the link between farming family members and farm tasks, but they have diminished the degree to which family members are exclusively involved in farm tasks.

Of course, the degree of involvement varies with the age of the family member. As shown in table 4.6, the share of male family members engaged principally in farm tasks only exceeds the share engaged primarily in non-farm tasks at ages 60 and over. The pattern for women is different. Many female family members in their twenties work outside agriculture, but the share of females engaged principally in farm tasks begins rising in the late twenties. By their forties, more women are working principally in agriculture than are not. At all ages up to 60, with the exception of groups under the age of 25 (when women hold full-time salaried jobs before marriage and childbirth), the number of female family members engaged exclusively in agriculture far outnumbers the number of men.

Table 4.6. Employment Situation of Farm Family Members by Age-group (for 1980)

Age (years)	Members working only on family farm	Members working on family farm and in non-farm job: Members for whom family farm is main occupation	Members working on family farm and in non-farm job: Other job is main occupation	Members working only in non-farm job	Members not working in a job
Men					
16	21.0	0.3	7.7	8.1	62.9
20–	10.6	2.1	42.4	32.3	12.7
25–	9.2	3.8	57.6	28.0	1.5
30–	10.7	5.5	64.7	18.0	1.0
35–	12.3	7.9	68.0	10.9	0.9
40–	15.4	11.0	66.1	6.8	0.8
45–	18.2	12.6	63.5	4.9	0.9
50–	22.7	12.9	59.0	4.2	1.1
55–	29.9	12.6	51.9	3.8	1.9
60–	45.3	10.6	37.0	3.0	4.1
65–	57.2	7.5	23.7	2.7	8.9
70–	52.3	2.4	6.7	1.6	37.0
Total	25.1	7.4	45.6	10.5	11.4
Women					
16	14.5	0.0	4.0	9.7	71.8
20–	11.7	0.4	24.0	46.7	17.2
25–	31.8	1.5	25.2	24.9	16.6
30–	41.3	3.5	29.7	13.2	12.3
35–	42.8	6.5	36.9	7.8	6.0
40–	43.7	8.6	39.2	4.9	3.6
45–	49.9	8.3	35.2	3.5	3.2
50–	59.8	6.6	26.5	2.7	4.4
55–	66.6	4.8	18.9	2.3	7.4
60–	70.4	3.0	10.9	1.8	13.9
65–	66.1	1.6	5.5	1.4	25.4
70–	35.4	0.4	1.1	0.6	62.5
Total	44.8	3.7	20.5	9.1	21.9

Source: *Nōgyō sensasu* [Agricultural Census].

These statistics give a general picture of the average employment situation for members of farming families. Older members and women perform the routine farm work. Men of mature age employed in non-farm jobs assist with farm tasks in a supplementary role. These men carry the principal burden for their families' livelihoods and work mostly in cyclical or seasonal

Table 4.7. Proportion of Fishing Family Members Employed in Family Business by Age, 1963 and 1978

Age (years)	Men				Women			
	Members over 15 yrs (A)	Members employed in family fishing business[a]			Members over 15 yrs (C)	Members employed in family fishing business[a]		
		Number (B)	%	$\frac{B}{A}$ (%)		Number (D)	%	$\frac{D}{C}$ (%)
1963								
15–	58,954	16,031	4.7	27.2	53,455	2,429	2.8	4.5
20–	100,942	60,922	17.9	60.4	95,980	17,367	19.8	18.1
30–	100,730	85,368	25.1	84.7	98,046	28,161	32.1	28.7
40–	66,333	60,175	17.7	90.7	76,855	20,558	23.4	26.7
50–	69,058	60,807	17.9	88.1	73,530	12,781	14.6	17.4
60–	95,942	57,021	16.8	59.4	94,073	6,467	7.4	6.9
Total	491,959	340,324	100.0	69.2	491,939	87,763	100.0	17.8
1978								
15–	43,876	4,677	2.0	10.7	40,580	313	0.4	0.8
20–	33,158	11,156	4.7	33.6	28,774	1,327	1.8	4.6
25–	32,978	15,485	6.5	47.0	26,578	3,751	5.0	14.1
30–	23,560	14,633	6.1	62.1	22,897	6,056	8.0	26.4
35–	24,206	18,686	7.8	77.2	29,700	10,497	13.9	35.3
40–	35,468	30,510	12.8	86.0	37,479	14,041	18.6	37.5
45–	42,486	37,837	15.8	89.1	38,261	13,963	18.5	36.5
50–	36,625	32,948	13.8	90.0	35,210	11,106	14.7	31.5
55–	24,628	22,058	9.2	89.6	28,488	7,103	9.4	24.9
60–	22,881	19,583	8.2	85.6	23,722	3,851	5.1	16.2
65–	56,589	31,597	13.2	55.8	63,641	3,604	4.8	5.7
Total	376,455	239,170	100.0	63.5	375,330	75,612	100.0	20.1

a. "Family members employed in family fishing business" refers to persons working for more than 30 days annually at sea in the family-run fishing business, and does not indicate whether the family fishing business or some other job is that person's principal occupation.

Source: *Gyogyō sensasu* [Fisheries Census].

wage-based jobs. As they grow older, management of their own farm becomes their principal occupation again for a time before retirement.

The situation in fishing families is similarly described in table 4.7. Note that the statistics are based on persons who spend more than 30 days annually at sea working to support their own fishing business.

Table 4.7 displays several features of persons employed in fishing tasks at sea. In the first place, men far outnumber women. More than 60 per cent of male family members participate in fishing tasks, compared to only 20 per cent for females. Those ratios changed noticeably, however, between 1963 and 1978, declining for men by 6 per cent and increasing for women by 2 per cent.

Meanwhile, there has been a decline in the number of both men and women working in the fisheries sector since the rapid economic growth period began. Paralleling the relative increase in the share of fishing tasks performed by women, as noted above, this drop in total numbers is pronounced among men, while the relative share of women in the total fishing labour force has risen. (During the period of this study, the rate of decline in the number of men was 29.7 per cent and in the number of women 13.8 per cent.)

A fourth feature observable in table 4.7 concerns the shares of different age-groups in fisheries among family members. More than half of all male family members over the age of 30 are engaged in fishing tasks, with peaks of around 90 per cent for those in their forties and fifties. The proportion of female family members is between 20 and 30 per cent for those in their thirties, forties, and fifties. Participation by women of all other ages drops off dramatically. The figures in table 4.7 also suggest that a large number of men in their twenties are employed outside the fisheries sector and that this tendency has grown even stronger, while women before marriage and of childbearing age take almost no role in fishing tasks at sea, and the age of retirement is low.

The share of different age-groups in the fishing labour force shows a marked trend between the years 1963 and 1978 toward a higher median age. During this period, the proportion of men over 50 in the total number of male family members employed in family-run fishing businesses rose from 34.7 to 44.4 per cent, and of women of the same age from 22.0 to 34.0 per cent. The percentage of men over 40 jumped from 52.4 to 73.0 per cent, and of women from 45.5 to 71.1 per cent.

Table 4.8 shows the proportions of fishing family members by their principal form of employment. Although the number of both men and women whose principal occupation is in the fisheries sector (sea-based) is fewer than the number of people engaged in fisheries tasks at sea for more than 30 days annually (see table 4.7), the difference is not large. In fact, more than 80 per cent of the latter group make fishing their principal occupation. In the broadest terms, then, we can say that persons engaged in fisheries tasks at sea make such tasks their principal occupation. A second feature to be observed in table 4.8 is the large number of women not engaged in fisheries tasks at sea but performing related shore-based tasks. The combined total of women in their thirties, forties, and fifties who work either in fisheries tasks at sea or in related shore-based tasks equals roughly half of all female family members.

Principal employment in non-fisheries-related occupations is concentrated among younger age-groups, particularly men in their twenties and women in their early twenties. Only a few men begin working in the fisheries sector after graduating from high school. The majority of young men take a wage-based job before joining the family fishing business when the head of the family is ageing. Likewise, few young women work in the business of their natural family before marriage. In general, a job in a company will precede marriage.

The data lead to the conclusion that the shift of young people to employment outside family-run fishing businesses has created a high median age and a constant increase in the percentage of women in this sector. At the same time, however, family members with their own fishing businesses display a strong tendency to continue fishing as their principal occupation.

The changes in employment patterns outlined above are reflected in the gradual change in the division of labour within families. In the immediate post-war period, two generations of males and several close male relatives would perform the tasks at sea, while women and older persons handled related on-shore jobs related to fisheries and/or farming. Today, the male head of the family or a husband-and-wife team performs tasks at sea. The family's male heirs follow wage-based jobs until a certain age; daughters take part-time jobs and at the same time contribute to the family's business in some way. The large size and mechanization of fishing vessels and their on-board equipment have reduced the number of workers needed at sea, and have permitted older men and women to perform tasks at sea. These factors are the same as those prevailing in the agricultural sector.

Responding to these developments, cooperative movement activity sustained by men in the prime of their lives and enthusiastic about owner-management of farm and fishing businesses has changed greatly under Japan's post-war democratic development. In farming, the majority of mature males have moved into non-farm jobs, with the result that decision-making in agricultural cooperatives has come to be concentrated in fewer hands (active farmers and coop employees). The limited advance of women into the management of cooperatives has not offset the heightening of this concentration. In fishermen's cooperatives, on the other hand, there has been a rise in the median age of members, but the high rate of participation by middle-aged men has resulted in greater membership authority than that in agricultural cooperatives.

The discussion above provides the fundamental economic details of transformed agriculture and fisheries in Japan during the rapid economic growth period. The following section outlines how the similarities and differences in the changes in these two sectors were related to their patterns of technological innovation.

Table 4.8. Principal Occupations of Fishing Family Members by Age-group, 1978 (percentages)[a]

		Self-employed					Employees			
			Fisheries							
Age (years)	Total	Subtotal	Sea-based	Shore-based	Agriculture	Other sector	Subtotal	Fisheries	Non-fisheries	Not employed
Men										
15–	27.5	13.8	10.3	1.9	0.7	0.9	13.8	3.3	10.5	72.5
20–	88.8	35.0	28.7	0.9	2.0	3.5	53.8	9.7	44.1	11.2
25–	97.0	47.5	38.9	0.7	2.7	5.2	49.5	9.9	39.6	3.0
30–	98.4	60.6	50.1	0.7	3.6	6.3	37.8	8.1	29.7	1.6
35–	99.2	72.4	62.6	0.5	3.9	5.4	26.8	7.6	19.2	0.8
40–	99.4	80.4	70.6	0.5	4.9	4.4	19.1	5.9	13.2	0.6
45–	99.7	82.9	72.5	0.6	5.6	4.2	16.7	5.0	11.7	0.3
50–	99.4	84.5	72.7	1.2	6.4	4.2	14.9	4.0	10.9	0.6
55–	99.0	85.9	71.4	1.7	8.2	4.6	13.1	3.4	9.7	1.0
60–	97.5	87.5	70.0	3.5	9.3	4.6	10.0	3.0	7.0	2.5
65–	76.4	72.5	48.1	10.3	10.5	3.5	3.9	1.2	2.7	23.6
Total	86.2	64.3	52.4	2.5	5.4	4.0	22.0	5.2	16.8	13.8
Women										
15–	18.6	6.5	0.7	4.1	0.6	0.9	12.1	0.1	12.1	81.4
20–	76.1	22.5	4.0	10.9	3.3	4.3	53.7	0.3	53.3	23.9
25–	77.3	49.2	12.5	21.2	8.5	7.0	28.1	0.7	27.4	22.7
30–	82.0	63.7	23.1	22.4	10.3	7.9	18.3	1.2	17.1	18.0
35–	87.6	69.5	29.7	20.7	11.4	7.9	18.0	1.3	16.7	12.4

40–	89.4	71.8	30.7	20.0	13.4	7.7	17.7	1.3	16.4	10.6
45–	88.8	73.2	29.2	20.6	15.6	7.9	15.6	1.1	14.5	11.2
50–	86.9	73.7	24.9	23.4	17.4	8.1	13.2	1.0	12.2	13.1
55–	82.9	72.6	19.3	25.7	19.6	8.0	10.3	0.9	9.3	17.1
60–	74.5	68.3	12.3	27.9	21.2	6.7	6.2	0.7	5.5	25.5
65–	42.8	41.1	4.3	18.5	14.5	3.7	1.8	0.2	1.5	57.2
Total	69.7	53.6	16.4	18.9	12.3	6.0	16.0	0.8	15.3	30.3

a. Figures show the percentage of the total number of persons in each age-group.
Source: *Gyogyō sensasu* [Fisheries Census].

II. The Effects of Technological Change

1. Agriculture

Technological innovation has developed in many aspects of agriculture, but nowhere more than in the labour-saving technology for rice cultivation. Rice yield has risen spectacularly from an annual tonnage of 10 million to 14 million. This is the result of dramatic reductions in labour requirements and a large rise in the productivity of the land. A great variety of technological advances lies behind these trends: improvement in soils, better irrigation, improved seed types, widespread use of pesticides, and greater use of fertilizers and mechanized equipment. In this section we will examine the relationship between this technological progress and the changes that have taken place in agricultural labour and economic patterns.

Table 4.9 shows that the time per 10-are plot required for the principal tasks of rice cultivation has markedly declined. Some difference is evident in the timing of this decline for particular tasks. The time needed for preparing fields for planting, and for irrigation and related tasks, dropped in the 1950s and early 1960s; that for planting, harvesting, and threshing in the late 1960s and after. The former period corresponds to the spread of powered ploughs, tractors, and irrigation pumps, the latter to the spread of rice-planting machines, harvesters, and combines.

Men are the principal operators of agricultural machinery. The general trend for men to undertake non-farm jobs jointly with their farm tasks has led to an increase in the number of women who operate farm machinery,

Table 4.9. Time Devoted to Rice Cultivation (per 10-are plot, 1952, 1965, 1975)

Area under cultivation	Year	Total labour time	Time spent on principal labour				
			Ploughing, raking	Planting	Weeding	Irrigation and related tasks	Harvesting and threshing
Average	1952	196.1	28.9	27.6	35.7	18.5	59.3
	1965	141.2	14.2	24.4	17.4	12.0	47.9
	1975	81.5	9.2	12.2	8.4	9.9	21.8
30–50 ares	1952	226.0	33.1	32.7	37.8	26.8	66.0
	1965	156.2	16.3	27.3	17.0	14.3	51.8
	1975	107.8	12.5	16.7	9.5	13.6	31.9
3 ha and over	1952	159.1	18.7	20.5	38.4	9.6	49.5
	1965	117.0	6.9	21.6	21.6	7.8	38.9
	1975	55.4	4.3	9.3	8.4	6.4	11.1

Source: Nōsei Chōsa Iinkai [Agricultural Policy Study Committee] supervised by Kayō Nobufumi, *Kaitei Nippon nōgyō kiso tōkei* [Revised Basic Statistics on Japanese Agriculture], pp. 488–489.

Table 4.10. Rate of Farm Expenses by Size of Farm, 1950–1980[a]

Area	Year	Gross farm income (unit: 1,000 yen)	Farm expenses (unit: 1,000 yen)	B/A (%)
Average for all farms	1950	189	46	24.3
	1960	342	129	37.6
	1970	937	450	48.1
	1980	2,275	1,370	60.2
Farms with less than 0.5 ha under cultivation	1950	75	19	25.6
	1960	137	57	41.6
	1970	275	149	54.0
	1980	558	408	73.1
Farms with less than 2.0 ha under cultivation	1950	406	102	25.1
	1960	946	339	35.8
	1970	2,287	999	43.7
	1980	6,619	3,785	57.2

a. These figures do not include those for farms in Hokkaido.
Source: *Nōka keizai chōsa* [Survey of Farming Family Economic Status].

but the prevalence of men in dealing with this task continues. The need for flexibility in performing farming tasks among men with commitments to non-farm jobs has led to a spread in individual ownership of machinery instead of joint use of machinery.

The mechanization of rice cultivation, together with the shift from self-supplied to purchased fertilizer, has contributed to a large rise in expenses. As shown in table 4.10, between 1950 and 1980 the ratio of farm expenses to gross farm income leapt from 24 to 60 per cent. The need to pay for machinery and fertilizer has severely limited the increase in net farm income.

Naturally, the above trend varies considerably with area under cultivation. Table 4.9 shows clearly that the labour time required per unit of land falls as the area under cultivation increases, and rises as that area decreases. It is also possible to infer from table 4.10 that expenses as a share of gross farm income decline in proportion to the area under cultivation.

Technological progress has eliminated the excessive labour of the high season for most farmers, and has allowed members of farming families to hold regular wage-based jobs. In fact, the low income generated by rice cultivation and the high costs of mechanization have heightened the need to hold a non-farm job. On the other hand, for the minority of large-scale rice-cultivators, technological advances have offered the potential to secure the income necessary to boost labour productivity to a far higher level than is possible on small farming units.

In running the majority of Japan's farms, women and older people now play the central role, while the few large-scale farms that exist are mainly run by middle-aged men. These men have expanded their operating base by renting land owned by others or by accepting commissions to cultivate separately owned land. Such expansionary trends face limitations imposed by rising land prices, however, and will not be able to eliminate small farming units through absorption in the foreseeable future.

A schematic description of the relationship between the employment patterns of members of farming families and the size of family-run farms shows three basic elements:

1. In large-scale operations, both men and women, including those in young age-groups, are centred on agriculture.
2. In medium-scale operations, men and young people are oriented toward non-farm occupations, while women and older persons are oriented to agriculture.
3. In small-scale operations, both men and women are centred on non-farm occupations.

This background illustrates the relative rise in the position of the labour of women in supporting agricultural tasks. None the less, the decline in the time required for rice cultivation has meant that female members of farming families have increasingly taken up non-farm occupations (see table 4.5). This tendency corresponds to growth in the part-time market in farming villages.

2. Fisheries

Technological innovation in coastal fisheries has displayed great diversity for each type of fishery activity. In the discussion below, we examine three such areas—trawling, breeding, and diving—which have shown contrasting patterns of technological change.

(1) Trawling

Trawling technologies have advanced in such areas as fishing vessels, navigational equipment, on-board equipment, and fishing tackle. Motors have been added to vessels, reducing the physical requirements of labour. Wood has given way to reinforced plastic, encouraging standardization and improved performance and durability of vessels. More powerful engines have speeded up navigation times, expanding accessible fishing grounds and securing fresher produce. Greater power has also raised trawling capacity. Larger vessels have enhanced the safety and comfort of fishing operations, and made it possible to operate even under adverse weather conditions or for continuous periods extending over several days.

Many of the navigational aids and on-board instruments used formerly only on large, deep-sea fishing vessels or cargo vessels have appeared on coastal fishing vessels in more compact versions. Foremost among navigational aids is wireless equipment; ancillary on-board instruments include fish-school detectors and on-board equipment directly used for fishing tasks

Table 4.11. Number of Workers at Sea at Height of Fishing Season (family members and hired hands, 1953–1978)

	1953		1963		1978	
Operating vessel	Family members	Hired hands	Family members	Hired hands	Family members	Hired hands
<3	2.0	0.9	1.7	0.2	1.4	0.0
3–5 tons	1.8	4.4	1.9	1.2	1.6	0.3
5–10 tons	1.6	8.7	1.8	3.7	1.9	0.9

Source: *Gyogyō sensasu* [Fisheries Census].

includes mechanized net-haulers and automatic fishing lines. This equipment has reduced the personnel requirement for navigation and fishing, and the level of skill needed for such tasks.

In parallel with the growth in the size of vessels, fishing tackle such as nets has become larger. Net materials have switched from cotton to synthetic fibres, offering greater strength and eliminating deterioration, thereby cutting onshore repair tasks. Technological improvements such as these have permitted the participation of older persons and women in fisheries tasks by lowering the physical requirements of labour and promoting efficiency.

As mentioned earlier, these advances did not significantly alter the total volume of the coastal fisheries catch, which remained constant at around 2 million tons. This limit to the catch is beyond human control—it depends on the rate at which marine life regenerates, and is a major characteristic of the fisheries industry. Faced with this limitation, participants in the industry were forced to compete with each other for a greater share of the catch of high-value fish. This competition fuelled the rate at which technological advances were adopted in the coastal fisheries industry. Naturally, the introduction of advanced techniques and equipment increased the level of expenses relative to the size of catches.

Table 4.11 shows the decrease in the size of crews on different classes of fishing vessels. As crews grew smaller, more labourers from the fishing industry sought higher wages in shore-based industries. This process led to reorganization into family-based operations, even among the top stratum of coastal fishing families.

(2) Marine Farming

Marine farming is an industry in which natural resources are reproduced through human management aimed at stabilization and promotion of production. It actually resembles land farming in several respects: specific areas of the sea are used exclusively for cultivation, and marine farmers have a pre-existing right of ownership of the harvest and assume the expenses of feeding and care. The problems of marine farming also resemble those of land farming: rapid increases in yield due to technological progress can lead to a slump in prices. We will consider these matters in the context of laver (*nori*) cultivation, the most important of all marine farming activities.

Progress in laver-farming technology produced new methods of propagation, as well as work procedures in the sea and on land. The advances in propagation are central, forming the basis of advances in other areas. The principal new developments are described below:

1. The long-established method of propagation had relied on the natural adhesion of naturally produced laver spores to bamboo frames. A better understanding of the growth process of the laver seaweed led to a shift to artificially cultivated spores. This eliminated variations in yield due to inconsistent quality of seedlings, and permitted hybriding and other improvements in the quality of laver. For the first time, it became possible to cultivate laver as an agricultural product.
2. In place of bamboo frames to which the spores attach and the laver grows, special nets were adopted. These nets greatly facilitated harvesting and other seaside tasks as well as permitting more extensive cultivation. The use of bamboo limited the area of cultivation to sea depths shallower than the length of the poles (several metres). Nets, however, could be floated at any depth, eliminating this limitation. The "free-floating" method, in fact, has opened up many new areas of the sea for commercial purposes.
3. The technique of freezing the laver nets to which seaweed shoots have adhered and exchanging nets during the cultivation season (October to April) has boosted yield by permitting two or even three crops per year. Harvesting at the optimum stage of growth has also overcome the problem of deterioration in quality as the laver ages.

Statistics clearly show the rapid growth in laver output due to these technological advances. Annual output of 1 billion sheets around 1950 had risen to almost 4 billion by 1960. With the widespread use of advanced techniques by 1970, output was up to 6 billion sheets. The peak of 9.6 billion sheets recorded in 1973 led to a sharp drop in prices, and since then annual production has been held at around 8 billion sheets to support prices. Initially, rising production increased the number of laver cultivation operations, but as techniques advanced further, prices fell and costs escalated.

Older laver-producing regions were pushed out of the industry as new regions flourished using the new technologies. These developments caused smaller businesses to move away from laver cultivation. Table 4.12 indicates the rise in the number of businesses involved in laver cultivation during the 1960s, and the abrupt drop in this number after that time. This turnaround is a vivid reflection of the rapid pace of technological change just described.

The growth in production of laver naturally led to increased labour requirements for harvesting and drying, promoting in turn the development of new technologies for these operations. Manual harvesting of seabeds gave way to the use of mechanical harvesters and the introduction of vessels de-

Table 4.12. Percentage of Businesses Involved in Laver Cultivation as a Primary Activity, 1963–1978

Year	Businesses engaged in laver cultivation (A)	Businesses with laver cultivation as primary activity (B)	Ratio of $\frac{B}{A}$ (%)
1963	58,620	51,354	87.6
1968	93,569	52,644	56.3
1973	46,094	38,436	83.4
1978	29,793	24,913	83.6

Source: *Gyogyō sensus* [Fisheries Census].

signed for laver-farming. Shoreside tasks before mechanization consisted of fine-chopping of harvested laver, spreading it in thin layers on drying sheets, removing the dried laver, and bundling. Laver was sun dried and therefore dependent on weather conditions.

Drying machines were introduced, eliminating the need to wait for good weather, and equipment gradually increased in capacity and functions until automatic drying machines appeared that integrated all steps up to the final bundling. Eventually, mechanization advanced to the extent that the harvested laver could be simply fed into a machine which turns out the finished product. Technological progress of this type is the principal factor behind the tenfold increase in output of laver over a short period. It also allowed family-run operations to establish a commercial business structure without having to acquire non-family labour.

Higher levels of technology, however, also demanded higher levels of capital investment. This need forced many operators to retire from laver cultivation. Among family operations that continued to cultivate laver, wider discrepancies arose in commercial viability and some families face continual economic difficulties.

Let us now describe the employment conditions in family laver-cultivation businesses. During the growing season from October to April, father-and-son or husband-and-wife teams work at sea, while older female family members perform shoreside tasks. The work involves long working hours over consecutive days. In the off-season, the growers prepare for the next season, and in many cases the women work in agricultural or part-time jobs, and men take work in the fishing industry. The percentage of women in laver-cultivating families who work at sea is far higher than in other branches of the fishing industry. The principal reason for this is that the work involves shorter distances from shore and less time, so that women are better able to meet the physical demands and balance the work with their domestic responsibilities.

Even when women do not take part in work at sea, they participate in shoreside tasks as a matter of course. Consequently, if we consider both the labour at sea and that on shore, we find that the majority of female members of families cultivating laver are involved in some way in the family

Table 4.13. Extent of Joint Agricultural Operations in Laver-farming Families, 1963–1978

	Year	Individual proprietorships (A)	Households with farmland No. (B)	Households with farmland ≥0.5 ha (C)	$\frac{B}{A}$ (%)	$\frac{C}{A}$ (%)
Laver-farming families[a]	1963	51,299	36,811	21,560	71.8	42.0
	1968	52,448	33,058	19,614	63.0	37.4
	1973	37,652	20,945	12,121	55.6	32.2
	1978	24,341	13,011	7,764	53.5	31.9
Individual proprietorships	1963	262,518	170,932	59,801	65.1	22.8
	1968	248,323	124,982	50,109	50.3	20.2
	1973	224,968	96,991	37,783	43.1	16.8
	1978	210,123	77,685	29,769	37.1	14.2

a. "Laver-farming families" refers not to all families involved in laver cultivation, but to those for which laver cultivation is the primary business. The data under "laver-farming families" in this table differ from those in table 4.12 because the latter includes businesses other than individual proprietorships (corporations and fishermen's coops).

Source: *Gyogyō sensasu* [Fisheries Census].

business. This situation means that it is difficult to secure a stable income throughout the year, and necessitates dependence on seasonal jobs with low stability. In many cases young women (daughters) engage in other industries to secure a stable income. Women of middle and advanced age are hired to replace them.

At present, a high percentage of households in the laver-cultivation sector also jointly engage in agriculture. As shown in table 4.13, only slightly more than one-third of all families in the fisheries sector engage in agriculture, but this percentage rises to over 50 per cent for laver-farming families. The percentage of such families cultivating more than 0.5 hectares of land is also higher—32 per cent, compared to 14 per cent for the fisheries sector as a whole. This is mainly due to the fact that laver-farming families engage in agriculture as a source of employment during the summer, which is the off-season for laver production. Another reason is that, in the early stage of technological advancement in laver farming, fishing families whose female members ran farms of a limited size as a family operation were able to begin cultivating laver. On the other hand, many fishing families whose female members were committed to wage-based jobs were unable to enter the laver-cultivation business.

(3) Diving

This branch of the fisheries industry is composed of women who dive to collect stationary shellfish such as abalone. Two features distinguish this occupation: the divers are solely women, known as *ama*, and technological progress has not led to the replacement of traditional, human skills

Technological advances in diving, including use of oxygen tanks and wet suits, have popularized it as a sport among the general populace but have not been adopted by *ama*, whose only equipment is a knife to pry shellfish from rocks. The main reason behind the failure of new technologies to appear in this branch of fishery is the slow pace of regeneration of shellfish resources. Abalone require from four to six years to grow, and any sudden increase in the amount collected would create a serious shortage in availability.

To prevent such shortages, the authorities stipulate overall general regulations, and fisheries cooperatives and the *ama* themselves abide by self-imposed restrictions. These regulations limit the diving season and the number of divers for specific zones, and as a general rule prohibit the adoption of new techniques.

Restrictions such as these run contrary to modern concepts of "freedom of enterprise," but the *ama* have strictly observed these rules and methods for hundreds of years. Furthermore, the method of passing on these specialized diving skills from one generation to the next has safeguarded the traditional rules from being infringed by outsiders.

In areas where shellfish resources exist but are not worked by traditional *ama*, permission to dive using modern equipment is accompanied by careful supervision to prevent the danger of exhausting resources by competitive operations. For example, a fisheries cooperative may be allowed to work in such a region using one diver and modern equipment, with the proviso that operations will cease should the danger of exhaustion of resources arise. The determined refusal of *ama* to adopt new techniques is a manifestation of their concern to preserve the resources, even as they compete with each other for higher catches.[3]

Ama follow two patterns when diving. They may dive alone, using a floating tub to which they cling to rest and into which they place their catch. Or they may work in a husband-and-wife team, the husband operating a boat while the wife dives. In this case, the *ama* uses weights manipulated from the boat by her husband to prolong her dive and ensure safety. These typical patterns are practised with variations in different regions.

III. Patterns of Female Labour in Family-run Fishing Businesses

1. Conditions of Female Labour

(1) Age

Patterns of female employment are significantly determined by age. Nowadays, it is the exception, rather than the rule, for a young woman to enter the family-run fishing business after graduation from high school or college (with the exception of shore-based laver cultivation). Normally, young women find jobs with a company, and return to work in a family-run business only after marriage. This trend is strengthened by the desire of many young women from fishing families to live in the cities. As a consequence, the sons of family-run fishing businesses often marry women from other backgrounds who have no prior experience of fishing. For these and other reasons, young women do not contribute to the family fishing business if other stable employment opportunities exist. During their childbearing years, young married women leave their wage-based jobs and devote themselves to housework and child-raising tasks, while at the same time contributing to the family's shoreside or agricultural activities.

After young children have grown, women find it difficult to re-enter the full-time labour force, so they tend instead to move into part-time occupations in family-run fishing or farming businesses or similar jobs. More women in their thirties and forties work in agriculture than in any other sector (table 4.7). Despite advances in mechanization and technology, however, work at sea in the fishing industry is more demanding than agricultural labour, and does not offer the flexible working hours of agriculture. In regular fishing activities, once a vessel puts out to sea it will normally operate for a stretch of about eight hours. These circumstances reduce the number of women in their fifties working at sea and cause most women to retire once they reach 60 years of age. At this stage, most women will concentrate on caring for their grandchildren and helping with domestic chores, though they will still assist with shore-based fishing activities and agricultural tasks.

(2) Composition of Households

In common with the agricultural sector, the size of fishing families has been diminishing since Japan's rapid growth period. According to the *Gyogyō keizai chōsa* [Economic Survey of the Fishing Industry], the average size of fishing families has declined from 6.8 persons in 1952 to 6.2 in 1960, 5.0 in 1970, and 4.3 in 1980. This decline is due both to families' having fewer children and to the tendency toward nuclear families. Two types of nuclear families exist—aged couples whose children have left the village, and middle-aged families who live separately from their parents. Both types are on the increase.

Older women in the former type of nuclear family are unable to work at sea, so that jobs of this type become the sole responsibility of older men. Older men, however, cannot physically match the performance of younger men, so their involvement in work at sea is limited to lighter tasks in fishing grounds near shore.

The absence of older women in the other type of nuclear family prevents an arrangement in which older women take responsibility for domestic and child-rearing tasks while younger women work on boats or in other jobs. Women are therefore unable to engage in full-time employment. Furthermore, the irregular schedules demanded of family-run fishing businesses, involving as they do nighttime work, prevents women from taking advantage of welfare facilities such as day care and preschools. These circumstances have strengthened the tendency for women to work in regular, wage-based daytime jobs.

In traditional fishing villages with a low level of urbanization, multigeneration households are still the norm. In such households, older women perform domestic and child-minding tasks, giving younger women a wider selection of employment options. The undeveloped labour markets in such areas, however, mean that young women tend to work at sea if there are employment opportunities in the fishing industry, such as during the laver-growing season.

(3) Development of the Labour Market

The extent of development of a region's labour market naturally has a big impact on employment patterns for women. Development of the labour market, however, occurs in different circumstances in the fisheries sector than in agriculture. If men shift to jobs outside the fishing industry, their families may be unable to support a family-run business because women alone cannot operate the business (except in the case of *ama* diving operations). This situation forces women, too, to seek employment outside the fishing industry. In areas with an undeveloped labour market and a long off-season, many men are forced to seek employment elsewhere for that period, as is the case in agriculture. During this season, women cannot support the fishing business alone and so their labour is either unused or diverted to extremely unprofitable tasks.

(4) Parallel Businesses

Seasonal variations in the fishing industry mean that labour requirements differ considerably between the on- and off-seasons. Fishing households whose womenfolk work in semi-permanent wage-based jobs can depend only on their regular labour resources even during the height of the season. On the other hand, female fishing-household members working in family-run businesses such as agriculture, retail selling or service operations can participate in fishing during the busy season, making possible an expansion in the scope of fishing operations. The benefits of this latter arrangement are most pronounced in those types of fishing operations that involve household-run processing or that require extensive periods of shoreside preparation. These benefits inhibit the women in families performing such types of fishing from taking wage-based jobs in favour of a parallel family-run business that leaves them free to participate during the busy fishing season, even if they are underemployed at other times of the year. The predominance of parallel farming businesses among families who cultivate laver (table 4.13) reflects a desire to keep the labour of female members close at hand in order to cope with the busy harvest period.

(5) Type of Fishing Activity and Composition of Crews

The type of fishing activity a household performs has a strong influence on female labour patterns. Women are needed to perform shore-based tasks if such tasks are a large component of overall activities (as in the case of processing of the catch and marine farming). Regular types of fishing by net or line do not require any shoreside tasks in particular.

The composition of the workforce at sea is also influential. Of coastal fishing families operating boats of less than 10 tons, 54 per cent rely on a crew of one, 33 per cent on two, and 13 per cent on a crew of three or more (based on 1978 Fisheries Census figures). Boats operated by a single crew member do not produce a large enough catch to require a large amount of shore-based work by women. The influence of two-person crews on female labour patterns depends on whether those two people are father and son, brothers, or a husband and wife. A man-and-wife team operating at sea will also share the shoreside tasks, thereby reducing the time available for work at sea. Male family members working at sea will catch that much more than a single-person boat, and will be able to remain at sea for longer than a husband-and-wife team. Consequently, the amount of shoreside work will increase proportionately, and since the male family members working at sea will have less time for shoreside tasks, female members will bear the principal burden of those jobs.

In addition, of course, there are many coincidental conditions and subjective preferences of the people involved, and these, too, influence labour patterns.

2. Fisheries Work and Female Labour

Patterns of female employment in the coastal fishing industry are largely determined by particular types of fishing. Table 4.14 indicates the number of men and women involved in various fishing tasks (offshore). The first observation that can be made is that more than 40 per cent of all women work in the marine farming industry and that women account for more than one-third of the total labour force in this industry. Laver cultivation is typical; women account for about 40 per cent of the at-sea workforce in this sector. Second, we note that one-fourth of all female workers are employed in the combined shellfish- and kelp-gathering sectors (including diving by *ama*). The ratio of women in the total workforce in each of the two sectors is more or less equivalent to that in the marine farming industry.

Table 4.14 shows clearly that marine farming and shellfish- and kelp-gathering are the principal work of women. These activities fit in well with the requirement that women be available for domestic tasks. They are performed close to shore. They involve relatively long periods of shoreside labour and subsequently require only brief periods of work at sea. Furthermore, women performing these activities are able to enjoy a reasonable degree of flexibility when they work. Finally, with the exception of *ama* diving, these activities do not require particularly demanding skills, so that women can easily substitute when the number of men decreases.

Small-scale fixed shore netting resembles breeding and gathering tasks. Schools of fish caught by fixed shore netting do move, but the area where

Table 4.14. Self-employed Persons in Fishing by Type of Work, 1978

	Men		Women		Ratio of women,
	Number (A)	Proportion (%)	Number (B)	Proportion (%)	$\frac{B}{(A+B)}$ (%)
Marine farming	58,693	24.5	32,022	42.4	35.3
Laver-cultivation	29,690	12.4	19,065	25.2	39.1
Shellfish-gathering	17,325	7.2	10,661	14.1	38.1
Kelp-gathering	14,658	6.1	7,993	10.6	35.3
Gill-netting[a]	30,940	12.9	8,437	11.2	21.4
Small-scale trawling	16,405	6.9	3,929	5.2	19.3
Small-scale fixed shore netting	7,039	2.9	2,112	2.8	23.1
Line-fishing[b]	39,336	16.4	2,753	3.6	6.5
Other	54,774	22.9	7,705	10.2	12.3
Total	239,170	100.0	75,612	100.0	24.0

a. Drift-netting of salmon and trout is not included in the figure for gill-netting.
b. Line-fishing of bonito, mackerel, and cuttlefish is not included in the figure for line-fishing.
Source: *Gyogyō sensasu* [Fisheries Census].

humans work is fixed by the position of the nets, so that the labour requirements are virtually the same as in breeding and gathering. The low number of family-run businesses working at fixed shore netting is responsible for the limited number of women working in this sector. When considered as a share of the total employment in this sector, the number of women is high—about one-fourth.

In contrast, strong limitations affect the extent to which women work in the line-fishing sector, as represented by work on fishing vessels. The catch in line fishing is highly mobile and the movements of schools of fish make it impossible for operators to determine their working schedules arbitrarily. The fishing grounds are distant from shore, making it physically difficult to work for only short periods. And line-fishing is not economically viable. In consequence, the component of women in the line-fishing workforce is a mere 6 per cent.

Of course, circumstances prevent the total exclusion of women from crews working on vessels, even in some fishing sectors requiring a high degree of work at sea. Women are included in crews for gill-netting, in which much time is spent removing fish from the nets. Women also participate in trawling to sort the catch by variety and size. Nowadays, the overall decline in the number of workers has led to more men working alone in these sectors without calling on the labour of women. In such cases, however, the tasks of removing fish from the nets and sorting must be performed either onshore (leading to an inevitable decline in the freshness and, therefore, value of the catch) or during breaks in the fishing process while still at sea. The direct economic advantages of these arrangements have meant that more women (mostly the wife of the man working the fishing vessel) will work on board despite the extra effort required to cover domestic tasks as well. Many fishing localities retain their aversion to the practice of women working at sea on board fishing vessels. But the desire to increase the level of the catch is proving stronger than that aversion.

3. Patterns of Female Employment in Coastal Fishing Families

This section reconsiders the patterns of female employment described in tables 4.7 and 4.8 on the basis of the above discussion.

(1) Continuous Work at Sea

This pattern refers to women who work on board fishing vessels continuously. Specifically, it includes women who work year-round at regular fishing tasks, and women who combine seasonal work in marine farming with work at sea. The inability, even today, of women to work alone means that this pattern usually involves a husband-and-wife team. Women following this pattern are not able, of course, to engage in wage-based work. This pattern, including cases of parallel work in agriculture, is the most self-contained form of family-run business.

(2) Seasonal Work at Sea

The most typical form of this pattern, in which women only work at sea on a seasonal basis, is marine farming, but the pattern also covers cases of regular fishing, when women work on board vessels only during the busiest periods, and of *ama* diving. Three household arrangements predominate in this pattern:

1. The combination of men working year-round at fishing and women working at agricultural tasks, part-time wage-based jobs, or shore-based fishing tasks during the summer, and at marine farming during the winter.
2. The combination of men working year-round at fishing and women working in tasks at sea in a supplementary role during the busiest season, and at agricultural tasks, part-time wage-based jobs, or shore-based fishing tasks as their regular occupation. Another pattern exists in regions such as the northern sector of the Japan Sea coast, where severe conditions make operation during winter difficult.
3. Performance of fishing tasks by husband-and-wife teams during the summer, and local or distant employment in wage-based jobs by the husband and wife (or husband alone) during the winter.

In all cases, fishing families give priority to their fishing business, with the result that any secondary activities in agriculture receive only perfunctory care, and wage-based employment is of a short-term, unstable character.

(3) Fisheries-related Shore-based Tasks as Principal Employment

Women in fishing families perform related shore-based tasks to differing degrees. When the volume of such work is low, the weight of work in other industries is high, but when the extent of shore-based-fisheries-related activities is high, women tend to pursue such employment as their principal occupation. Three arrangements predominate in this pattern:

1. The head of the family business and his son crew the fishing vessel, thereby raising the size of the catch over that of a single-person operation (in this case, women take responsibility for preparing fishing tackle, for coiling ropes, and for landing the catch).
2. Family-operated processing of the catch, thereby earning income from value added (this covers tasks such as drying of kelp, removing shellfish from their shells, and boiling).
3. Door-to-door sales of the catch by women in areas where a regional market exists.

Under these arrangements, women work as long as, and frequently longer than, men working at sea, also taking responsibility for domestic tasks. These circumstances virtually exclude any opportunity for entering the wage-based labour market.

(4) Parallel Employment in Other Sectors as Principal Occupation with Secondary Employment in Fisheries-related Shore-based Tasks

When the scale of a family-run fishing business is small and the size of the catch too is small, even women performing shore-based fisheries-related tasks are able to find wage-based employment in non-fisheries sectors. Often, such employment is also a matter of necessity. Such women, however, work extensively in the family fishing business during the busy season, and also tend to assist routinely when fishing vessels are leaving or entering port. In consequence, the pattern of this wage-based employment is strongly influenced by the irregular schedule of a family-run fishing operation. It is based on the short term and is changeable both seasonally and daily. Women working in this pattern are employed at the lowest levels of pay, mostly as part-time labour in processing fisheries produce. Operators of processing firms adapt to these conditions by having detailed knowledge of the lifestyle patterns among family-run fishing businesses.

(5) Regular Wage-based Employment

In regions close to urban centres where employment can be obtained easily, many young and middle-aged women work as employees in companies, the civil service, or fisheries cooperatives. Jobs of a full-time nature in supermarkets and offices also fall in this category. Young women who do not yet have children (particularly those who have just graduated) commonly hold such jobs. But because the levels of pay in these jobs far exceed those in temporary positions, some older women who have children also opt to work full-time, depending on nurseries or older family members for child-minding tasks. Many women not from a fishing background, but who married into fishing families, seek jobs in which they can use skills acquired in urban employment before marriage. Despite these conditions, on the whole many young women encounter difficulty in resuming jobs they held before marriage, childbirth, and child-raising when faced with stiff competition from new graduates.

The fixed hours of such regular employment virtually exclude any contribution to the shore-based tasks of the family-run fishing business. One result is that a shortage of workers to perform these tasks may occur during the busy season. Different families find different solutions. Older family members will sometimes fill the gap. Sometimes the family's male members will shorten their hours at sea to cover shore-based jobs. Or housewives from nearby farming families may be hired. In such situations, women will continue to work in their full-time jobs as long as their income is higher than the loss incurred through a decline in the size of the catch, or than the costs of hiring outside help.

The female labour market in farming and fishing villages has developed rather extensively around a focus on part-time employment. This development has occurred in the localized industry for processing farm and fisheries produce, in garment-making factories, and in factories that subcontract the assembly of machine components, as well as in various supporting home-based jobs. Striking regional variations are evident in this development, however. It is far advanced in villages located near centres of regional commerce, but seriously retarded in outlying regions where employment opportunities are almost zero. Nationwide factors stimulate further development: the regional relocation of factories spurred by continuing rapid economic growth; the expansion of demand for labour with the growth of small regional industries; greater commuter convenience with better roads and higher levels of motorization; and the improved supply of labour as the use of labour-saving devices has spread in homes and as social welfare facilities such as day care and nurseries have increased.

Even with more numerous employment opportunities, however, poor conditions of labour (low rates of pay and insecure tenure of employment) do not offer enough incentive for women to abandon their family-run businesses. Consequently, most women are forced to double participation in the family-run business with a wage-based job, and employers have to adapt holidays and working hours to fit in with family-run businesses.

These considerations limit the extent to which women in farming and fishing families can perform wage-based jobs, and even the nature of these jobs. The practice of holding a parallel wage-based job has given women in farming and fishing families an idea of the wage value of their labour in the family business, thereby strengthening their desire to put the family business on a rational commercial basis that reflects the worth of the labour involved. And, as women have moved out of domestic isolation, they have secured a greater sense of independence through earning individual incomes of their own. This is another factor that has promoted modernization of the social relations in Japan's farming and fishing villages since the end of the Second World War.

IV. Conclusion

Technological progress was slow to start in Japan's farming and fishing industries before the war. Despite this, the female members of farming and fishing families moved into a wide range of jobs. Struggling to subsist in the face of poverty, the underdeveloped nature of labour markets forced these women to accept employment patterns with a high level of servitude to their employers. These conditions have changed dramatically since the war, offering women the option to choose where they want to work by comparing a range of possibilities.

Labour markets in cities and in farming and fishing villages have expanded. The nature of family-run businesses has changed in response to technological progress. These movements have presented women in farming and fishing families with new problems and new choices. Do they leave their village for the city after graduating from school, or do they stay? Do they want to marry into a farming or fishing family, or not? How will they divide up jobs among family members? What job suits them best? And as they grow older, new questions confront them.

The way individual women answer these questions is shaping employment patterns. This chapter has attempted to generalize about these answers and to uncover those patterns.

Notes

1. This chapter does not consider capital-based fishing operations run using hired employees, and concentrates only on family-run businesses. The former include deep-sea and offshore fishing, the latter coastal fishing involving vessels of 10 tons and under.
2. Except when running farming business at the same time, jobs held jointly by the head of a fishing family are limited virtually to seasonal employment at times when no fishing is taking place. The only jobs that can possibly be performed jointly during the fishing season are coastal activities such as kelp-gathering. Such jobs generate only a tiny income.
3. The refusal to adopt modern diving techniques leads inevitably to accidental deaths, such as when a diver runs out of breath while on the ocean floor. *Ama* divers are not the only examples of fisherpeople who depend on the regeneration of natural resources and who oppose the introduction of available modern techniques. A common means of fishing abalone in many regions is to pierce the abalone from a boat with a long pole fitted with a sharpened hook. Only this method is permitted in these regions, which also limit the size of motors fitted to boats.

Bibliography

Agricultural Administration Research Committee. *Chiiki nōgyō no tenkai toshufu nōgyō* [Development of Local Agriculture and Housewives' Agriculture]. *Nihon no nōgyō* series, no. 146. 1983.

Agriculture, Forestry, and Fisheries Research Council. *Sengo nōgyō gijutsu hattatsushi* [Post-war Development of Agricultural Technology]. Nōrin Tōkei Kyōkai, Tokyo, 1970. 10 vols.

Furushima Toshio, ed. *Sangyō kōzō henkakuka ni okeru inasaku no kōzō* [The Structure of Rice Cultivation under Industrial Structural Reforms]. Tōkyō Daigaku Shuppankai, Tokyo, 1975.

Hasegawa Akira, ed. *Nihon gyogyō no kōzō* [The Structure of Japanese Fishing]. Nōrin Tōkei Kyōkai, Tokyo, 1981.

Isobe Toshihiko et al., eds. *Nihon nōgyō no kōzō bunseki* [Structure Analyses of Japanese Agriculture]. Nōrin Tōkei Kyōkai, Tokyo, 1982.

Kaneda Teishi, ed. *Nihon gyogu gyohō zusetsu* [Illustrated Explanation of Japanese Fishing Tools and Methods]. Seizandō Shoten, Tokyo, 1977.

Nakayasu Sadako. *Nōgyō no seisan soshiki* [Production Organization of Agriculture]. Ie-no-Hikari Kyōkai, Tokyo, 1978.

Saga University, Faculty of Agriculture. *Nori yōshokugyō no keizai bunseki* [Economic Analyses of *Nori* Cultivation]. Written by Jinnai Yoshito. 1979.

Segawa Kiyoko. *Ama* [Women Divers]. Miraisha, Tokyo, 1970.

Takayama Ryūzō et al., eds. *Gendai suisan keizairon* [Economic Studies of Fisheries Today]. Hokuto Shobō, Tokyo, 1982.

Tanaka Noyo. *Ama-tachi no shiki* [Four Seasons for Women Divers]. Shinjuku Shobō, Tokyo, 1983.

Yagi Tsuneo. Nori yōshokugyō no keiei bunseki [Analyses of *Nori* Cultivation Business]. *Suisan keizai kenkyū* (Fishery Agency), no. 34 (1981).

Chapter———5

Innovation and Change in the Rapid Economic Growth Period

Sakiko Shioda

In 1983 there were 22.63 million women in the Japanese labour force, 14.86 million working as salaried employees. Compared to the figures for 1955—17 million in the labour force, 5.31 million as salaried employees—these figures represent increases of approximately 1.3 and 2.8 times respectively. With the corresponding rates of increase for male workers over the same period being 1.5 times and 2.2 times, it is evident that, over the past nearly 30 years, the numbers of female salaried workers have increased significantly.[1] The typical attributes of the female salaried worker have also changed considerably, from the pre-war pattern of young, single women working on a short-term basis, primarily in the textile industry, to women working on a long-term basis, continuing even after marriage, and middle-aged and older women returning to the workforce after temporarily taking time off for marriage and childbearing.

This trend is likely to continue as employment patterns diversify and the service sector grows. With housewives leaving the home to work, home-making, child-rearing, and the care of the elderly and sick are being taken over by public welfare and commercial services. At the same time, the traditional role distinctions, with men working outside and women staying in the home, are becoming blurred. The greater participation of women in the labour force is affecting more than simply the workplace and management practices, and is also having a profound effect upon the family and welfare services.

In Japan, women first entered the wage-earning labour force in significant numbers during the post-war period of rapid economic growth[2] during which major technological advances were made. This period began in 1955 and lasted until the 1973 oil crisis. Automation of the production of light electrical appliances and precision instruments, the introduction of office automation, and the expansion of the sales sector all combined to broaden the opportunities available to women workers, who had until then been mostly confined to the textile industry. Such advances also created a new demand for the latent workforce of married women. Not only labour-saving automation in the workplace, but also a new affluence and change in life-styles generated by the introduction of many new technologies, were critical factors in stimulating this change in the Japanese female workforce.

Between 1965 and 1970, at the peak of the rapid-growth period, nearly every home had running water, electricity, and gas utilities, and electric rice-cookers, washing machines, refrigerators, and vacuum-cleaners were commonplace. These electrical appliances, along with increasingly available processed and instant foods, freed housewives from many time-consuming tasks. The decrease in the birth rate, extended life-spans, and higher education for women all combined to create extra time for women. The availability of more time, as well as the increasing need to supplement the family income, turned many Japanese women's interests away from the home to work outside.

Today, rapid advances in micro-electronics technology are affecting all industrial sectors, in particular contributing to an expansion in services, and this is bringing about more significant changes in female labour patterns. In view of these changes, let us look back at the post-war years of rapid economic growth, the changing nature of salaried employment at that time, and the new issues which emerged for the working woman as a result of these changes.

I. Changes in the Employment Structure

1. Increase in Salaried Female Workers

The acute shortage of labour generated by the spurt in economic growth was met primarily by siphoning off labour from the farms. This led to a reduction in female labour in the primary industries that was greater than the corresponding reduction in male labour, and at the same time brought about a greater increase in the salaried female workforce than in salaried male workers.

Over the 1960–1970 decade, at the peak of Japan's rapid economic growth, male labour in the primary industries of agriculture, forestry, and fisheries dropped from 25.8 to 20.0 per cent and female labour fell sharply from 43.1 to 26.2 per cent. In contrast, the number of workers in the secondary and tertiary industries increased (table 5.1). The same pronounced increase in female workers as compared to male workers is evident when we look at the two decades following 1955, our base year. Over this time-span the number of male workers nearly doubled as opposed to a 2.2-fold increase in female salaried workers (table 5.2).

As shown in table 5.3, there was little change in the percentage of self-employed workers among the total female labour population during this time, but the proportion of women working for family-owned businesses (including farming and forestry) dropped while that of salaried workers

Table 5.1. Number of Workers by Industry, 1960–1980 (10,000 persons)

		Industry[a]		
Year	Total number	Primary	Secondary	Tertiary
1960	M. 2,660 (100.0)[b]	686 (25.8)	931 (35.0)	1,042 (39.2)
	F. 1,712 (100.0)	738 (43.1)	345 (20.2)	628 (36.7)
1965	M. 2,902 (100.0)	570 (19.6)	1,061 (36.6)	1,270 (43.8)
	F. 1,861 (100.0)	604 (32.5)	429 (23.1)	826 (44.4)
1970	M. 3,172 (100.0)	475 (20.0)	1,241 (39.1)	1,456 (45.9)
	F. 2,039 (100.0)	534 (26.2)	530 (26.0)	974 (47.8)
1975	M. 3,338 (100.0)	376 (11.3)	1,300 (38.9)	1,661 (49.8)
	F. 1,964 (100.0)	361 (18.4)	505 (25.7)	1,093 (55.7)
1980	M. 3,385 (100.0)	294 (8.7)	1,321 (39.0)	1,770 (52.3)
	F. 2,142 (100.0)	283 (13.2)	605 (28.2)	1,250 (58.4)

a. Primary industries refers to agriculture, forestry, fisheries, and mining; secondary industries to construction and manufacturing; and tertiary industries to wholesale marketing, retail marketing, finance, transportation, communication, utilities (electricity, gas, water), services, and public services.

b. Numbers in parentheses are percentages.

Sources: Prime Minister's Office, *Kokusei chōsa* [National Census], 1960–1975, and *Rōdōryoku chōsa* [Labour Force Surveys].

Table 5.2. Number of Salaried Workers by Sex, 1945–1980

	Total (10,000 persons)			Index (1955 = 100)	
Year	Male	Female	% women	Male	Female
1945	152	71	22.0	12	13
1950	962	342	26.2	84	68
1955	1,247	531	27.9	100	100
1960	1,632	738	31.1	131	139
1965	1,963	913	31.8	157	172
1970	2,210	1,096	33.2	177	206
1975	2,479	1,167	32.0	199	220
1980	2,617	1,354	34.1	210	255

Source: Prime Minister's Office, *Rōdōryoku chōsa* [Labour Force Surveys], except for figures for 1945, which are based on Ministry of Commerce and Industry, *Kōgyō tōkeihyō* [Industrial Statistics].

rose. In 1970, the percentage of salaried female workers exceeded 50 per cent of all female labour. This trend has continued through the 1973 oil crisis, marking the end of the rapid-growth period and the beginning of the period of slow economic growth. By 1980, the distribution of male labour had shifted to 64.3 per cent salaried, 24.1 per cent working for family-owned business, and 11.6 per cent self-employed.

Table 5.3. Total Female Working Population by Type of Occupation, 1960–1980

Year	Total	Salaried	Self-employed	Family business
Percentages				
1960	100.0	41.9	13.5	44.7
1965	100.0	49.2	12.1	38.6
1970	100.0	53.2	13.9	32.9
1975	100.0	59.7	12.0	28.1
1980	100.0	64.3	11.6	24.1
Fluctuations (1,000 persons)				
1960–65	1,483.0	1,977.1	−54.7	−470.1
1965–70	1,888.0	1,752.2	602.7	−435.6
1970–75	−883.1	813.4	−516.2	−1,227.6
1975–80	1,347.2	1,767.8	72.2	−460.7

Source: Statistics Bureau, Prime Minister's Office, *National Census*.

2. Married Women in the Workplace

The second major change to take place among female workers was the increase in middle-aged and older married women entering the labour market as salaried workers. This has changed the female labour participation curve from one that peaked at a young age bracket to an M-shaped double curve with two peaks, one for very young workers and one for middle-aged and older workers (fig. 5.1). Unlike in other countries, there is a conspicuous drop in female workers in the 25–35 age-group. This group is the one most likely to quit for marriage and childbearing.

Since the period of rapid economic growth, the proportion of very young female workers aged 19 and younger has dropped significantly, while that of female workers aged 40 and older has gone up (fig. 5.2). There are two reasons for this: the higher level of education among women today means that more young women are in school, while at the same time married women in their forties and older, whose children have grown up, are returning to the workforce.

A shortage of male workers compelled corporations to hire these middle-aged and older women. The effect by 1975 was a greater percentage of married female workers than of single female employees. By 1980, this percentage was up to 57.4 per cent, with more than half of all salaried female workers being married (fig. 5.3). This is quite a significant change from the pre-war years, when most female workers were very young and single.

Most of the married women entering the labour force at this time were employed at unskilled simple tasks on a part-time basis. The rate of increase among part-time women workers considerably exceeded that among men, and though the 1976 oil crisis temporarily put a damper on part-time

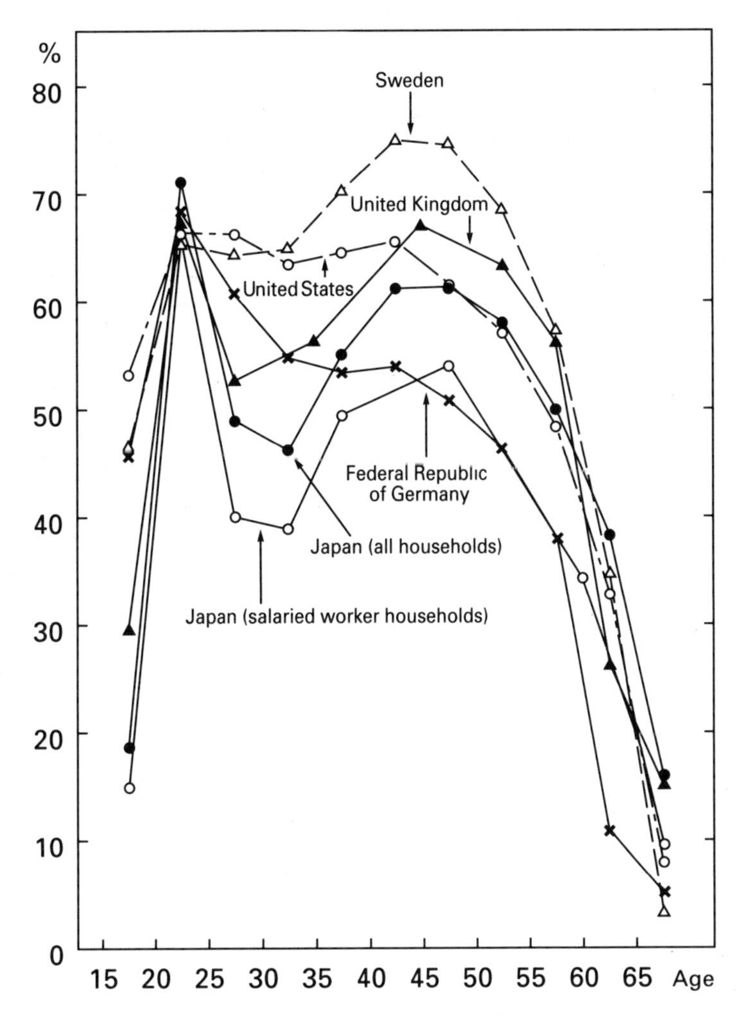

Fig. 5.1. Salaried Female Workers by Age, 1960–1980
Source: Prime Minister's Office, *Rōdōryoku chōsa* [Labour Force Surveys].

employment for women, their number continues to increase (fig. 5.4). In the two decades between 1960 and 1980, the proportion of part-time workers rose from 6.3 to 10.0 per cent of the total labour force. Among women, the increase was especially pronounced, going from 8.9 to 19.3 per cent. A 1980 survey[3] found that among married working women, the percentage of part-timers at 73.4 per cent was far above the 34.7 per cent of those working full-time.

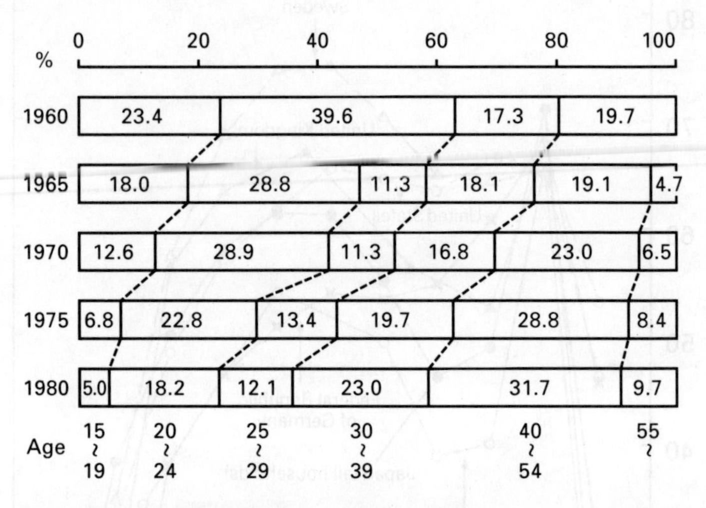

Fig. 5.2. Salaried Female Workers by Age, 1960–1980
Source: Prime Minister's Office, *Rōdōryoku chōsa* [Labour Force Surveys].

Year	Single	Married	Divorced or widowed
1962	55.2	32.7	12.0 %
1965	50.3	38.6	11.1
1970	48.3	41.4	10.3
1975	38.0	51.3	10.8
1980	32.5	57.4	10.0
1983	31.1	59.5	9.4

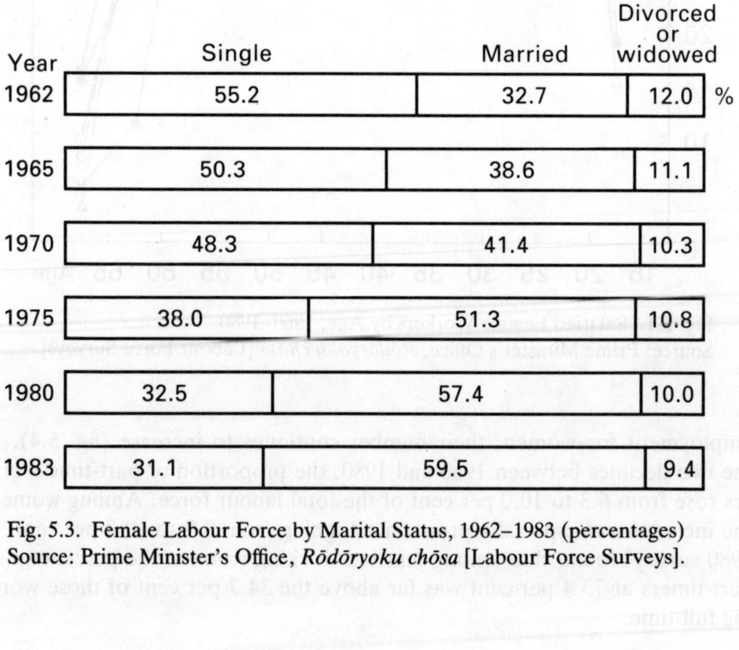

Fig. 5.3. Female Labour Force by Marital Status, 1962–1983 (percentages)
Source: Prime Minister's Office, *Rōdōryoku chōsa* [Labour Force Surveys].

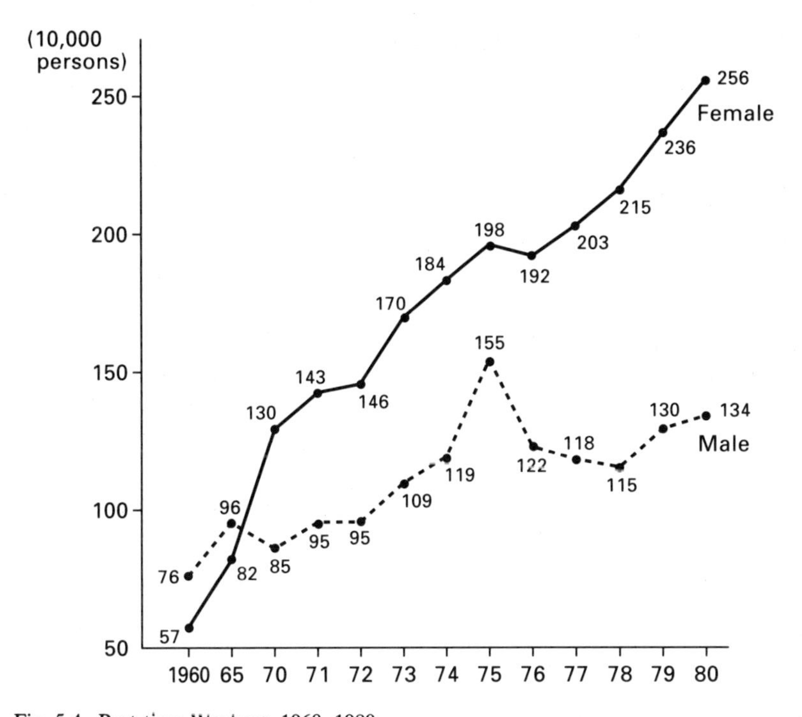

Fig. 5.4. Part-time Workers, 1960–1980
Note: Part-time workers are defined as those who work less than 35 hours per week (including seasonal workers and those working irregular hours). Figures for 1960 and 1975 have not been adjusted for time-series connections.
Source: Prime Minister's Office, *Rōdōryoku chōsa* [Labour Force Surveys].

3. Changes in Types of Work

The third change that has taken place in female labour in Japan has been the diversification of employment opportunities engendered by changes taking place in the industrial structure.

As shown in table 5.4, the occupations with the greatest number of workers and with the highest percentage of female workers in the 1960–1970 decade were: professional/technical, clerical, sales, skilled/manufacturing, physical labour, and services. There was a significant increase in the number of workers in management, transport, and communications during this decade, but though the number of women in these fields rose correspondingly, their percentage of the total never became very significant.

Clerical, sales, and skilled and manufacturing occupations were the most directly affected by technological advances. The introduction of automation into mass production actually caused clerical and management functions to increase and, of course, led to an increase in sales volume. Companies sent

Table 5.4. Number of Salaried Female Workers by Job Type, 1960–1980

Job type	1960	1965	1970	1960–70 rate of increase (%)
Professional/technical	60 (33.3)[a]	76 (37.6)	100 (40.7)	166.7
Management	2 (2.5)	4 (3.4)	5 (3.8)	250.0
Clerical	170 (35.9)	251 (39.9)	339 (46.9)	199.4
Sales	58 (34.7)	88 (37.0)	112 (32.6)	193.1
Agriculture/forestry/ fisheries	24 (32.9)	14 (23.7)	10 (23.8)	41.7
Mining/quarrying	2 (5.7)	1 (5.0)	1 (10.0)	−50.0
Transport/communica- tions	5 (5.3)	22 (12.0)	22 (10.0)	440.0
Skilled/manufacturing		220 (24.9)	291 (25.9)	
Physical labour	240 (26.9)	70 (31.5)	66 (33.2)	148.8
Services	108 (54.8)	150 (54.7)	150 (56.2)	138.9

a. Percentages (in parentheses) show proportion of women in total (male and female) workforce.

Source: Prime Minister's Office, *Rōdōryoku chōsa* [Labour Force Surveys].

their male workers out as salesmen and installed female workers in their new offices and retail outlets. A new type of female worker appeared on the scene, having at least a high-school or junior college education, quite capable of handling complex office work, and thoroughly versed in her company's product line.

The simplification of work processes in the manufacturing sector led to an increase in female workers here as well. A major change in the structure of the manufacturing sector further contributed to this increase. Prior to the era of rapid economic growth, female production workers were concentrated in the textile industry, but with the advent of rapid growth they shifted to the metal and machinery manufacturing industry. In 1954, women accounted for 43.1 per cent of all workers on textile production lines and only 14 per cent in the metal and machinery industries. By 1961, the shift from one sector to another was becoming apparent, with women comprising 32 per cent of all textile workers and 24 per cent of metal and machinery workers. In 1970 the percentages had completely reversed: 42 per cent in metal and machinery and only 18 per cent in textiles.[4]

With the rapid technological advances made during the rapid-growth years and the increase in automated manufacturing processes in general machinery, electrical appliances, and the petrochemical industries,[5] experienced male workers were replaced by unskilled female labour. The shift was directly related to the shift in export industries from textile to metal, machinery, and electrical appliances.

Another such shift has been taking place since the 1970s as micro-

Table 5.5. Salaried Female Workers by Job Type, 1960–1980 (percentages)

	Professional/ technical	Management	Clerical	Sales	Manufacturing	Services	Other	Total (10,000)
1960	8.6	0.3	24.5	8.3	34.5	15.5	8.3	100.0 (738)
1965	8.7	0.5	28.8	10.1	33.2	14.5	4.2	100.0 (913)
1970	9.1	0.5	30.9	10.2	32.6	13.7	3.0	100.0 (1,096)
1975	11.6	0.9	32.2	11.1	28.3	13.7	2.2	100.0 (1,167)
1980	13.0	0.8	32.7	11.6	27.2	12.9	1.8	100.0 (1,354)

Source: Prime Minister's Office, *Rōdōryoku chōsa* [Labour Force Surveys].

electronics and office automation stimulate a further change in Japan's industrial structure. The number of female employees in such tertiary industries as information services, research, advertising and publicity, retailing, insurance, and social services is increasing at a rate of 30 to 45 per cent. In the secondary industries, by contrast, while the rate of increase in female workers in precision machinery and electrical appliances is holding steady at around 30 per cent, their number is actually declining by 10 to 20 per cent in textiles, steel, paper and pulp, and petrochemicals, a rate of decline that is faster than that for men in these same industries.[6]

The story is not quite the same when it comes to professional and technical occupations and administrative management positions, however. A higher level of education and greater expectations have indeed fostered an increase in the number of women in these kinds of jobs, but, as of 1980, the percentage of all female workers in professional and technical occupations was only 13 per cent, and in management only 0.8 per cent (table 5.5). Even though the ratio of female workers in the workforce definitely went up during the rapid-growth era, this increase was primarily in the office, manufacturing, and service fields, and what professional work women did perform was still concentrated in jobs traditionally filled by women, such as child care (day care), elementary and junior high-school teaching, and nursing. Doctors, lawyers, university professors, and researchers are still primarily men, and these high-paying positions have yet to be seriously challenged by women.

For this it will be necessary for corporations to be more equal in their hiring and employment practices for men and women, and there will have to be an increase in social services and a change in perceptions of male and female roles if women are to be freed from the double burden of both outside employment and homemaking. It is notable in this context that, during the UN-declared "Decade for Women," many countries established equal opportunity laws reflecting the change that is taking place in the traditional perceptions of man as breadwinner and woman as homemaker.[7]

II. The Effects of Automation and Computerization

1. New Jobs for Women

The technological advances of the rapid-growth years were a major factor in opening up traditionally male-dominated jobs to women. According to a survey of 2,040 companies of all sizes and types,[8] approximately 30 per cent said they were using women in positions formerly reserved for men, the greatest rate of increase in their employment of women taking place between 1965 and 1967.

The larger corporations and factories producing electrical appliances, machinery, non-ferrous metals, metal products, steel, and chemicals (table 5.6) were the most aggressive in hiring women. All of these industries had a

Table 5.6. Secondary Industry Companies Providing New Job Types for Female Workers, 1968 (percentages)

Industries with many new job categories for female workers		Industries with few new job categories for female workers	
Chemicals	41.4	Foods	22.0
Steel	38.5	Tobacco	23.8
Non-ferrous metals	43.2	Textiles	17.9
Metal products	38.5	Lumber/wood products	19.2
Machines	49.0	Paper/paper products	25.4
Electrical appliances	50.4		

Source: Compiled from Women's Employment Office, Employment Promotion Project Corporation, *Joshi rōdōryoku no dōkō to joshi ni atarashiku hiraketa shokushu* [Trends in the Female Labour Force and New Job Categories] (1969), p. 20.

pronounced shortage of male workers in the past, and all had seen major advances in automation of their production and work processes. By installing new equipment and adopting improved safety measures, the large corporations were able to make up for the lack of male labour by hiring unskilled female workers. Small- and medium-sized businesses had always had more female workers than the large corporations, but among them also the shortage of male workers during the high-growth years led to an increase in the hiring of female workers.

According to a Ministry of Labour survey, 57.7 per cent of new positions for women were in manufacturing, 19.6 per cent in clerical and office work, and 13.8 per cent in completely new job categories created by the introduction of computers into the workplace. Looking at the reasons for hiring female workers, as shown in table 5.7, the most common reasons given by manufacturing companies were that automation had simplified work processes so that they could be handled by unskilled female workers and that certain aspects of jobs performed by male workers were being assigned to women. Non-manufacturing companies said they had found female workers better at certain jobs than men, and, again, that certain aspects of jobs performed by men were being relegated to women. As for professional and technical positions, quite a number of companies stated that they had given these positions to women on an experimental basis.

A 1969 survey showed that there was growing diversity in the job types being made available to women between 1963 and 1968 (table 5.8) in both manufacturing and non-manufacturing industries.[9] In the heavy machinery and chemical industries, automation and simplified and smaller equipment made it possible to have women workers operate lathes, drills, and mills, as well as take over washing and gas and electric-welding operations in shipbuilding and automobile production.

With the shortage of male technicians, women stepped in to take over

Table 5.7. Reasons for Employing Female Workers by Job Type, 1968—Multiple Responses (percentages)[a]

Reason	Profes-sional/ technical	Clerical	Skilled/ manufac-turing	Unskilled	Total
Female workers better at certain jobs than men	55.9	47.4	28.6	17.7	31.9
Women hired to make up for short-age of male workers	14.7	15.6	25.1	25.3	23.3
Women can be paid less than men	11.8	8.6	15.7	21.5	14.4
Automation has simplified processes so that they can be handled by women	5.9	9.3	47.3	41.8	38.8
Certain aspects of jobs performed by male workers assigned to women	47.1	45.4	47.0	40.5	45.9
Assigned on an experimental basis	32.4	7.3	11.7	10.1	11.5
Other	11.8	13.6	6.1	10.1	8.4
Total[b]	100.0 (34)	100.0 (303)	100.0 (1,499)	100.0 (79)	100.0 (1,915)

a. A few companies that gave no reason for hiring female workers are not included in this table.
b. Figures in parentheses indicate actual numbers.
Source: Ministry of Labour, Women and Minors Bureau, *Joshi rōdōdōsha no shūrō jōkyō no henka ni kansuru chōsa* [Survey on Changes in Types of Jobs for Female Workers] (1969).

such semi-skilled work as chemical analyses in the manufacturing industry and drafting in the non-manufacturing sectors. This was unprecedented, but it must be kept in mind that women filled only subsidiary positions, acting as assistants rather than becoming professionals or acquiring the necessary technical expertise themselves. Women have been hired in large numbers for the assembly of electrical appliances, thanks to the development of the conveyor-belt assembly system and the printed circuit board.

Mechanization has also opened up new jobs for women in the food and textile industries, sectors with a traditionally high percentage of female workers. For example, at one major food-processing company, there were 31 women working on automatic sausage-stuffing machines in 1963, 27 operating high-speed packaging equipment in 1965, and seven women in charge of automatic canning equipment in 1967. In the textile industry, the

Table 5.8. New Job Types for Women, 1963–1968

Manufacturing	Metalworker, lathe operator, metal press operator, mill operator, electrical appliance parts assembly, semiconductor element maker, light-bulb and vacuum-tube assembly, rolling-machine operator, wiring, casting-machine operator, welder, platemaker, transport machinery assembly, transport machinery inspector, optical machinery parts assembly, clock-assembly lens-grinder, automobile assembly, chemical analyst, drafter, finishing workman
Information processing	Keypuncher, checker, computer operator, teletypist, transcriber, telex operator, programmer
Management	Dormitory manager, section chief assistant, subsection chief, foreman or group leader, personnel placement director, assistant branch manager, branch office director

Source: Compiled from Women's Employment Office, Employment Promotion Corporation, *Joshi rōdōryoku no dōkō to joshi ni atarashiku hiraketa shokushu* [Trends in the Female Labour Force and New Job Categories] (1969), pp. 60–83.

introduction of automatic spinning machines, presses, cutting machinery, and dyeing processes has led to women replacing men in the operation of this kind of equipment.

Computers have contributed to startling advances for female workers in the non-manufacturing industries by opening up opportunities as keypunchers, computer operators, and other positions related to information-processing and data telecommunications. At first these kinds of jobs were limited to the finance and insurance, wholesale and retail, transport and communications, and utilities (electricity, gas, and water) industries, but between 1963 and 1968 they also became available in the chemical, steel, machinery, and electric appliance manufacturing industries.

Other jobs for women in the non-manufacturing sector were those related to the sales and marketing of new products developed during the rapid-growth years. These include, for example, new gas and electrical kitchen appliances, microwave ovens, and other home appliances. Women are now giving advice and providing technical services for these kinds of products, and they are considered to have an advantage in their better understanding of what the homemaker needs. It is in these areas that women have for the first time become involved in new product development and sales.

2. Working Conditions

How has automation affected working conditions for women, especially in the electrical appliance and precision instruments manufacturing industries and in computer-related jobs?

Replacing the textile industry in the rapid-growth period as the primary export industry, the electrical appliance industry offered new opportunities for young, single women workers. Whole groups of young women who had just completed their compulsory education (graduation from junior high school) were hired *en masse* by companies in the industry—a trend heralded with much fanfare in the media—but in actuality work in the modern, automated factory was not always as wonderful as it was made out to be.

Assembly of transistor radios, calculators, and tape-recorders on a line is tedious, repetitive work. At the Toshiba Yanagimachi plant, there was an almost complete turnover of assembly line workers at least once every three years, the reasons given for quitting not being marriage or childbirth, but the desire to be free of the monotony and tension of the work. Contributing stress factors were the shift-work system and the requirement that all workers live in company dormitories. Young female workers were not very happy with the restrictive dormitory life.[10]

Stress and high turnover rates were evident in other studies as well. An electric appliance factory in the Tokyo suburbs speeded up its assembly line, forcing workers to do their jobs faster. In just six months roughly 20 per cent of the women on the line quit, many of them to find new jobs in offices and coffee shops, later gravitating to bars and cabarets. The female workers at this particular company stated in response to a questionnaire prepared by the company's labour union that, after a day on the assembly line, everything seemed to keep moving even though working hours had ended.

In a study of the health problems of female workers in a television assembly plant, Sony discovered that the women who spent their day doing very fine wiring work and welding operations complained after three months or so of double vision, lack of appetite caused by fatigue, near-sightedness, astigmatism, burns, and rheumatism.[11]

At the peak of the rapid-growth era, a 1965 survey found that wages in manufacturing were especially low, second only to wages in the mining and construction industries (table 5.9). In the electrical appliance manufacturing industry, female workers were receiving above-average salaries, but still considerably less than males in the equally-booming steel industry; even though the gap between female and male wages was not as wide as in many other sectors, women were still only being paid 49.8 per cent of men's wages (table 5.10).

The results of a study conducted by the Ministry of Labour's Women and Minors Bureau revealed that many of the same features characterized working conditions for women in the precision instruments industry as in the electrical appliance industry.[12] This survey found that women involved in making measuring, medical, and optical equipment and timepieces shared four characteristics: (1) they were junior and senior high-school graduates and had taken up their occupation in place of work in the textile industry; (2) the majority were young and single with little work experience; (3) a considerable number lived with their parents or siblings and used their in-

Table 5.9. Women's Wages and Discrepancy with Men's Wages by Industry, 1965

Industry	Monthly wage (yen)	Percentage of male wage
Mining	18,173	41.4
Construction	19,372	45.3
Manufacturing	19,796	44.7
Production	18,371	47.5
Management	25,125	43.3
Wholesale and retail	22,939	50.0
Finance	32,393	47.5
Real estate	25,176	43.3
Transport and communications	30,310	60.4
Utilities	36,251	58.3
Total	22,275	47.8

Source: Ministry of Labour, *Maitsuki kinrō tōkei chōsa* [Monthly Labour Statistics].

Table 5.10. Women's Wages and Discrepancy with Men's Wages by Manufacturing Industry Sector, 1965

Industry	Women's monthly wage (yen)	Women's wages / Men's wages (%)
Manufacturing	19,796	44.7
Food-processing	18,249	40.5
Tobacco	39,884	67.5
Textiles	18,109	43.6
Clothing, textile products	16,256	43.0
Lumber, wood products	15,478	47.0
Furniture, interior products	17,131	50.8
Paper, pulp, and paper products	19,995	44.3
Publishing, printing, related industries	27,143	51.0
Chemicals	23,978	47.1
Petroleum and coal products	25,256	44.1
Rubber products	20,105	49.6
Leather and fur products	20,396	49.9
Ceramics and stone products	19,658	45.1
Steel	24,826	48.1
Non-ferrous metal	23,164	49.5
Metal products	20,044	50.7
General machinery	21,153	50.7
Electric appliances	19,830	49.8
Transportation equipment	23,187	50.0
Precision instruments	23,113	54.0

Source: Ministry of Labour, *Monthly Labour Statistics*.

comes freely, putting away money for marriage or to contribute to the family income; and (4) companies hired these women primarily because they were deemed skilled at detail work, were patient, persevering workers and could be paid low wages. These findings indicate that there was little change in female working conditions from the days when most female labour was concentrated in the textile industry.

Though many companies built modern, totally automated factories, the pressing need for rapid expansion of production capacity often compelled them to adapt old factories to new production processes. Many of these factories were old wartime munitions plants and spinning plants, their floors dirt or concrete and their lighting and ventilation very poor. The Ministry of Labour study cited above noted these conditions and urged that they be improved. A high near-70 per cent of women workers in such factories complained of ailments directly related to their work, mostly tired eyes and stiff shoulders.

In terms of wages, women in the precision instruments industry were also paid less than their male counterparts whom they had replaced at machining and inspection processes. The difference between men and women was also evident in compulsory retirement ages. Of the companies surveyed by the Ministry of Labour, roughly one-third had different retirement ages for men and women, most of them age 35 for women and 50–55 or no compulsory retirement age for men. These companies saw their female workers primarily as cheap, disposable labour.

During the rapid-growth era, computerized office work had as high an appeal among women as the electrical appliance industry. In actual fact, however, working hours were irregular and overtime was common. Poor working environments coupled with the repetitive and sedentary nature of the work caused many women to quit, and in some companies female workers were not expected to last very long.

Female keypunch operators, teletypists, and computer programmers stated that acquiring a skill in computer operation made it easy for them to change jobs, but at the same time they complained that little was known of how long hours at computer terminals might affect their health and that long working hours made it nearly impossible to continue their jobs after marriage.[13]

Computerization in financial institutions, beginning around 1960, led to a sharp increase in complaints by keypunchers and teletypists of physical and mental distress caused by their jobs. Their most common complaints were of pain in their hands, fingers, arms, backs, and shoulders, and eye fatigue. Some claimed that they could not even lift their arms to wash their faces and that they suffered from insomnia and restlessness. Very soon the same symptoms were appearing among women working as telephone operators, checkers (cash-register operators), assembly-line workers, and clerical workers. So common were these symptoms that they were given a name—tenosynovitis—and because most of these jobs were filled by women, female workers suffered the most. It was not long before the new disease became an issue of widespread concern.[14]

While technology was providing women with new opportunities as professionals and technical workers, it was also creating many new kinds of problems. Computers and automation made jobs simple, repetitive, and monotonous, a continuing trend today as micro-electronics permeate the business world. Already the terminal operation of video displays, a relatively simple process requiring little skill, is on the way to becoming women's work. If the kind of occupational health problems that appeared in the rapid-growth period are to be avoided this time, it will be necessary to create new work standards and to limit working hours. Women should also be given more opportunities, both within and outside the company, to train for positions in data-processing management and research and development.

III. Female Part-time Workers

1. Part-time Workers and Types of Work

Whether they are hired to work on a regular basis, only temporarily, or for just one day at a time, part-time workers are defined as those who work fewer hours per day, week, or month than full-time employees. During Japan's period of rapid economic growth, many middle-aged and older housewives took on part-time jobs and were called *pāto taimā* (part-timers).[16]

Part-time workers became most prolific in factories and offices in the 1960s as corporations sought to make up for the labour shortage by hiring housewives who had finished with early child-rearing. Technological advances and automation simplified work processes and made it possible to use unskilled labour. Part-time work enabled the housewife to earn an income without detracting from her duties as mother and homemaker, and she was generally satisfied if she earned just enough to supplement her husband's income and cover increasing household expenses.

At first it was assumed that "part-timers" were a temporary phenomenon proliferating in response to the labour shortage caused by the nation's rapid economic growth. Yet the number of part-time workers has continuously increased since the rapid-growth era, especially in tertiary industry, only decreasing temporarily when companies were forced by the oil crisis to streamline their operations. Even today there is no sign that the increase is temporary. Through the rapid-growth era to the present, part-time workers have continued to be an important source of labour. The average number of years part-time workers stay at one job has increased significantly over the nearly ten years between 1970 and 1979, increasing from 2.5 to 3.8 years in the manufacturing sector and from 2.1 to 3.2 years in wholesale and retail industry (fig. 5.5).

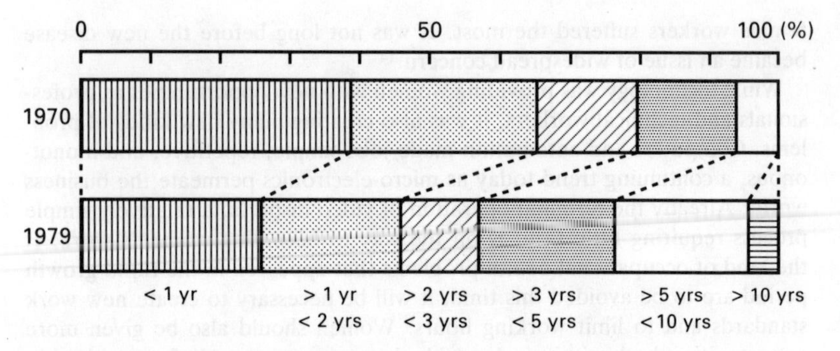

Fig. 5.5. Average Number of Years Female Part-time Workers Stay at One Job, 1970 and 1979

Sources: For 1970, Ministry of Labour, *Joshi pāto-taimu koyō chōsa* [Employment of Female Part-time Workers], 1971; for 1979, Ministry of Labour, *Daisanji sangyō koyō jittai chōsa* [Employment Conditions in Tertiary Industries] (see notes 17 and 20).

A 1970 Ministry of Labour survey found that part-time workers were concentrated in the manufacturing and wholesale and retail sectors, with 89.3 per cent of those in manufacturing involved in production processes compared to 73.9 per cent of those in wholesale and retail sales. At the time of the survey, most part-time workers were assigned unskilled jobs with low pay,[17] the most common being parts assembly, packaging, sales, money-collection, cleaning, cooking, and dishwashing.

A look at the part-time workers employed by Hitachi's Mobara plant gives a typical view of female part-time labour during the rapid-growth years.[18] This particular plant hired its first part-time female workers in the autumn of 1959, starting with some 30 to 40 housewives from farming households, to help with cleaning, shipment preparation, and transport. In 1968, wives of salaried company workers were added to bring the number up to 300, and part-time female workers were assigned throughout the plant, their duties expanding to include the assembly of vacuum tubes and electrodes for colour and black-and-white television sets. Hitachi had made the decision to hire more part-time workers after it found that it just could not maintain a staff of younger female workers capable of keeping up with its ever-increasing production rate. Once it had hired middle-aged and older housewives to fill out its labour force, Hitachi found that it had tapped an unexpectedly rich source of labour. This caused the company to reconsider its long-standing policy of hiring only young, single women, and to shift to a greater reliance on part-time labour made up primarily of older, married women.

Newspaper flyers proved an effective mean of recruiting part-time workers, as did simple word-of-mouth transmission among friends and neighbours. In 1968, the average age of the female part-time worker was 34.4, and though she usually had children, they were already of elementary-school age, freeing her for much of the day. Money earned through part-time work went primarily to pay school fees and to otherwise supplement the household budget.

To ensure that the part-time housewife labour force would continue working at its plant for an extended period of time, Hitachi first made sure that their husbands had no objection to their working. Special commuter buses were provided to make commuting as easy as possible, and within the plant part-time workers were treated much the same as their full-time counterparts despite differences in wages. As the Hitachi example demonstrates, the company's willingness to foster this new source of labour dovetailed nicely with housewives' desire for supplemental income and greater availability of time. Companies in the southern Kanto area and in the Kyoto-Osako-Kōbe industrial zone, where the shortage of young labour was especially acute, were among the first to hire such part-time workers on a large scale.

2. Improvement in Wages and Working Conditions for Part-time Workers

Female part-time workers quickly became a mainstay of the rapid-growth era, yet though they most often worked at the same kinds of jobs and for nearly the same hours as their full-time counterparts, they were not earning nearly as much. At one time this discrepancy was corrected somewhat, bringing the part-time hourly wage to approximately 82 per cent of that of the full-time worker in 1976. Yet in 1980 this percentage had actually dropped to 76 per cent. As female part-time workers became an established part of the labour market, the gap between their wages and those of full-time workers gradually widened, at the very least remaining stable and showing no signs of closing.[19]

As for working conditions other than wages, there does appear to be some improvement among the larger companies, though slight differences in the targets of surveys made in different years make accurate comparison difficult. According to a 1965 study,[20] 26 per cent of part-time workers were receiving social security benefits and 30 per cent health insurance benefits. By 1982, both figures had gone up to around 40 per cent. Only 9.6 per cent of the companies surveyed said they paid some kind of retirement allowance to their part-time workers. On the other hand, companies that paid bonuses to part-time as well as full-time workers had gone up between 1965 and 1982 from 35 to 62.7 per cent. Without the lifetime employment guarantee of the full-time employee, the part-time worker naturally found the bonus a greater incentive than a lump-sum retirement allowance.

With part-time workers accounting for a growing segment of the labour market and with companies finding more and more varied uses for such workers, it is inevitable that this will have a major effect on hiring practices for the full-time employee and on labour management policies. Considering this, it would seem logical that part-time workers be paid more, but a number of factors act to keep part-time wages low. Companies are reluctant to increase the wages of what they perceive as a cheap labour pool, and there are any number of middle-aged and older housewives seeking part-time jobs. Furthermore, many of these housewives do not want to earn more than the limit set by the government on non-taxable income earned by a dependent (as of 1985, the ceiling was 900,000 yen per year).

Another factor that has delayed the payment of better wages to part-time workers is the fact that they have not been organized to establish a strong bargaining position, and the labour unions already in existence have so far shown little interest in improving their plight. It is only very recently that the national labour federations in Japan have started to look into the possibilities for organizing part-time workers and to begin making serious efforts to seek improved wages.[21] The first such efforts took place in the distribution industry, particularly in supermarkets, which are major employers of part-time workers. Even here, however, the labour unions are motivated more by the fact that the number of full-time workers is dwindling, threatening their own continued existence.

Local unions took the lead, the Shōgyō Rōren publishing its "Standards for Part-time Employees" in June 1979 and the Japan Federation of Textile Industry Workers' Unions issuing a statement on "The Organizing of Part-time Workers" in May 1980. Among the major national labour federations, the General Council of Trades Unions of Japan (Sōhyō) in January 1981 and the Japanese Confederation of Labour (Dōmei) in June 1982 both announced their intention to organize part-time workers. Recognizing part-time work as a legitimate employment category, these national union federations have committed themselves to closing the remuneration and work-benefits gap between part-time and full-time workers. The Dōmei has made a distinction between those who wish to remain part-time workers and those who wish to become full-time employees, and is concentrating on improving the possibilities for part-time workers wishing to go full-time.

The motivation and types of occupations of part-time workers have diversified considerably since they first appeared on the scene in the era of rapid economic growth. As figure 5.6 shows, housewives seeking part-time work are coming from higher income brackets. The need to supplement the household budget is less pressing for them than is their desire to enjoy a more affluent lifestyle and to spend their free time constructively employed. With this new kind of labour and the greater importance being placed on workers' attitude and ability today, companies' employment and payment policies for part-time female workers will inevitably diversify according to types of jobs.

There are 3.06 million part-time female workers in Japan today, and the Ministry of Labour has finally got around to establishing certain minimal regulations concerning their employment.[22] Companies with ten or more part-time workers on their payroll must set employment regulations specifically for these workers. Part-time workers employed for 22 hours or more

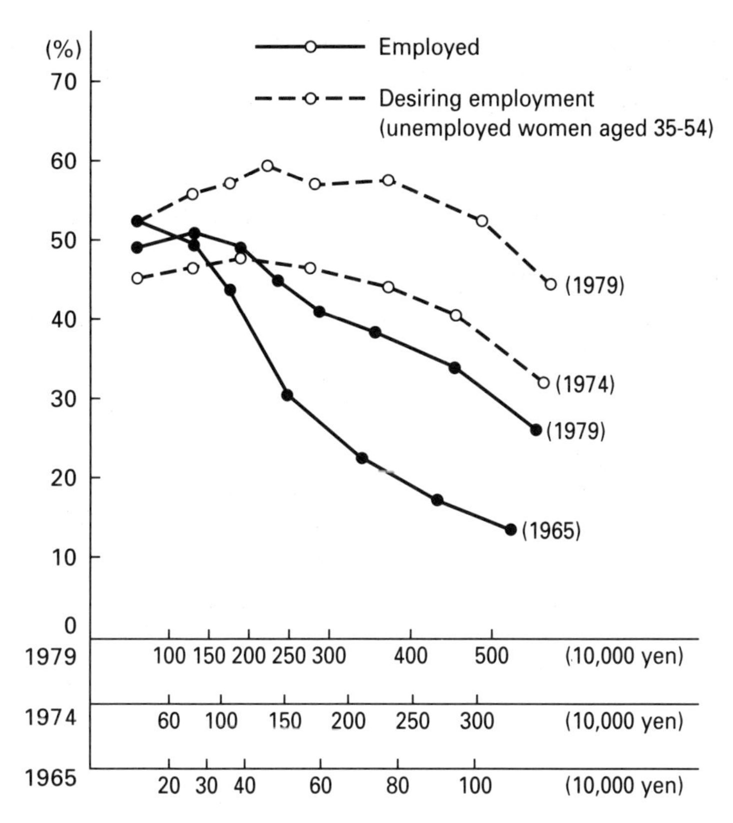

Fig. 5.6. Household Income Bracket of Married Women, Employed and Desiring Employment, 1965, 1974, and 1979
Source: Shioda Sakiko, "Kyōkyū-gawa kara mita saikin no joshi pāto-taimā no dōkō" [A Supply-side View of Recent Trends among Female Part-time Workers], in Koyō Shokugyō Sōgō Kenkyūjo [Employment Research Institute], *Koyō to shokugyō* [Employment and Jobs], vol. 42 (1982).

a week or who work at least three-quarters of the hours of a full-time employee are eligible for unemployment insurance. Those employed four days a week must be given some annual paid days off from work, and, finally, part-time workers who meet certain conditions must be paid salaries equivalent to their full-time counterparts.

Since they first emerged during the years of rapid economic growth, part-time workers have been a major force in Japan's economic development, and the housewife working on a part-time basis is likely to become a permanent fixture of Japanese society. Today, women looking for part-time positions are often highly educated and have specialized skills and previous work experience, and a growing number of them will probably be gravitating to professional and technical jobs. Companies would do well to tap this rich lode, not only improving the working conditions for conventional part-time positions but opening up more and better job categories to the part-time worker.

IV. Social Ramifications of Women in the Labour Force

1. Changes in the Home

The post-war period of rapid economic growth brought about major changes in the home as a result of the widespread availability of electricity, gas, and running water, and the many new products appearing on the market as a result of significant technological advances. Over the period 1956–1965, more than 60–70 per cent of all households had acquired an electric washing machine, electric rice-cooker, and electric refrigerator, and by the end of the rapid-growth era just about every home had these appliances (fig. 5.7). This, of course, greatly reduced the time required for household work, and led to the kind of major change in lifestyles indicated in figure 5.8.

Rice, the staple food, could be cooked with just the touch of a switch—no more need to chop wood and keep a cooking fire going. Laundry could

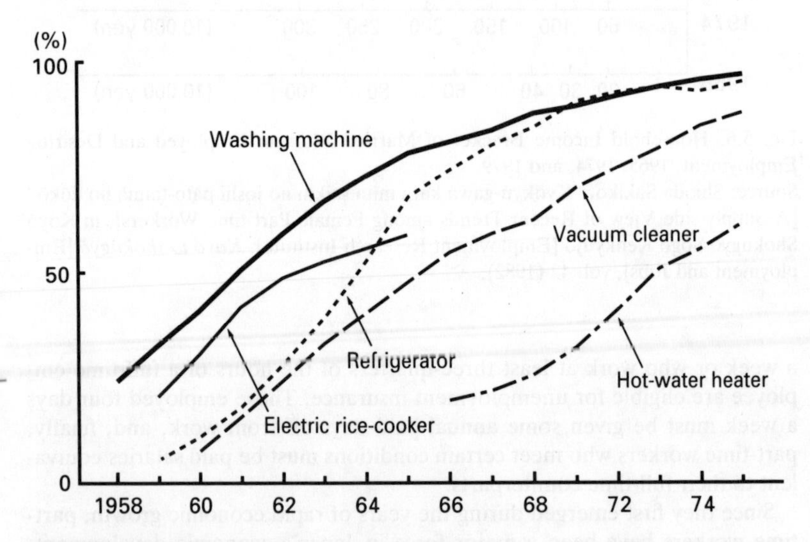

Fig. 5.7. Ownership of Electrical Household Appliances, 1958–1974
Note: Surveys were conducted in February of each year in cities with populations 50,000 and greater.
Source: Economic Planning Agency, *Shōhisha dōkō yosoku chōsa* [Forecasts of Consumer Trends].

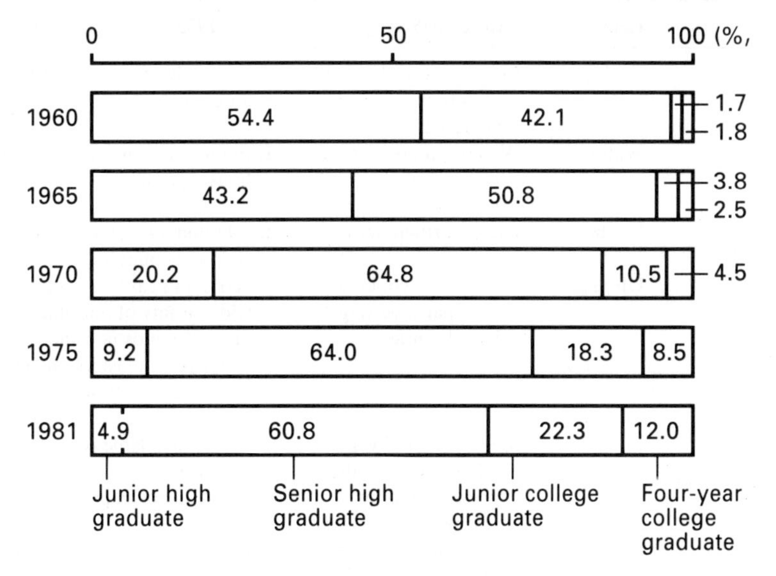

Fig. 5.8. Women Graduates by Educational Level, 1960–1981
Source: Ministry of Education, *Gakkō kihon chōsa* [Basic Survey of Schools].

be washed in a machine filled with water from a tap—no more need to haul up buckets of well water and spend backbreaking hours hand-washing each item of clothing. Meals could be prepared with processed, semi-processed, and frozen foods stored in a refrigerator. Inexpensive, ready-made clothing made it unnecessary for the housewife to spend late nights sewing and mending her family's wardrobe. When she did want to sew or knit, there were sewing-machines and knitting-machines that greatly speeded up the process.

All these new conveniences freed women, particularly housewives, from the strenuous labour that had once been necessary simply to keep the household going and to maintain the family's welfare. According to a Japan Broadcasting Corporation (NHK) 1975 survey of how people spend their time, the post-war housewife of a typical salaried worker's household spent approximately three hours less on household chores than her pre-war counterpart and had roughly four hours of free time to herself every day.

Life was certainly easier than it had once been, but the many new conveniences also meant a greater burden on the household budget. Their husbands' salaries not being enough to cover these extra expenses, housewives sought their own sources of income to supplement the household budget, and this is what generated the mass entry of married women into the labour market during the years of rapid economic growth. The necessary social services to support this influx of female labour were still far from sufficient, however, there being few child-care facilities, for example, or corporate

Table 5.11. Changes in Household Chores before and after 1955

	Item	Before 1955	After 1955
Food	Fuel	Cooking stove heated with firewood and charcoal	Gas and electric stoves with automatic pilot
	Water	Well or pump water or water hauled in buckets	Running water inside the house, gas hot-water heaters
	Kitchen	Dark, earthen floor	Bright and airy, stainless steel sink and counters
	Utensils	Iron pot for cooking rice, pans, cutting board, knife	Electric and gas rice-cookers, wide variety of pots and pans, thermos jars and hot-water pots, toasters, mixers, refrigerator with freezer, electric or gas oven
	Food	Meat, fish, and vegetables, homemade preserved foods (pickles and *tsukudani* soysauce-boiled preserves)	All foods available regardless of season Increased consumption of dairy and meat products, processed, semi-processed, and pre-cooked foods, variety of frozen foods.
Clothing	Laundry	Soap, well or pump water, washing by hand	Synthetic detergents, running water, electric washing machine, dry cleaning
	Fabrics	Cotton, hemp (natural fibres)	Natural and synthetic fabrics
	Apparel	Hand sewn and knitted, frequently mended	Ready made and made-to-order clothing Sewn on electric sewing and knitting machines
Housing	Furniture	Storage chests	Dressers, closets, couches and armchairs, carpeting, stereo, television, beds.
	Heating/ cooling	Handheld fans, *kotatsu* (quilt-covered table with charcoal brazier underneath), open fire pit, briquettes, open brazier	Electric *kotatsu*, kerosene and gas heaters and air-conditioners, electric fans, central heating
	Cleaning	Broom, duster, rags, dustpan	Vacuum cleaner, chemically treated dust cloths, mops, wax, cleaning solutions
Attitudes		Frugality: repair and recycling, tightly controlled budget, low living standard	Preference for disposable products, "consumption as virtue," used items readily replaced with new purchases, higher standard of living, increased spending and higher prices

Source: Tanabe Giichi, ed., *Katei keieigaku sōron* [Household Management] (Dōbun Shoin, 1977), adapted from p. 134.

policies accommodating the special needs of the working mother.[23] Most women, also, wanted only to work to the extent that they would still be able to fulfil their traditional functions of mother, wife, and homemaker, and because of this they opted for part-time rather than full-time jobs.

Thus, though there were now a greater number of women in the labour force, few had the careers or the working benefits of their male counterparts. Instead, most of these women worked for low pay at simple, unskilled tasks made available to them through advances in automation and mechanization.

2. Equal Employment Opportunities

In the democratization that took place just after the Second World War, the equality of men and women under the law was recognized for the first time, and equal educational opportunities provided for men and women. This, combined with the fact that people began to have more money to spend on education, resulted in a sharp increase in the number of women going on to senior high school and higher education. The two-year college entry rate for women increased from 3.0 per cent in 1960 to 11.2 per cent in 1970, and the percentage of women going on to four-year universities nearly tripled over the same period, jumping from 2.5 per cent in 1960 to 6.5 per cent in 1970. As is evident in figure 5.8, the educational level of women entering the labour force rose significantly over this decade, the percentage of those who had only gone as far as junior high school shrinking drastically.

The narrowing of the educational gap between men and women has contributed to more equal employment opportunities as well. Women, now just as highly educated as men, are increasingly looking for job opportunities once considered the exclusive territory of men. This was already evident in the rapid-growth period, as women began appearing in management positions traditionally assumed to be the exclusive preserve of men. Whereas there were only approximately 20,000 women in management positions throughout the country in 1960, their number had grown to roughly 110,000 by 1980. In terms of number, this is still insignificant, but in terms of growth it represents a hefty, more-than-fivefold, increase. The entry of women into traditional male-dominated job categories is certain to continue to escalate.

Another notable factor in the growing number of women in the labour force has been the major change that has taken place in women's lifestyles (fig. 5.9). In 1930, women typically married around age 22, had an average of five children, and died soon after the youngest had married. The pre-war Japanese woman devoted her whole life to husband, children, and often the husband's family-run business. In contrast, in 1974, the final year of the rapid-growth era, women were spending a longer time in school and therefore marrying later, having only two or three children at the most, and simply living longer. As a result, they had many years of free time ahead of them even after they had finished raising their children.

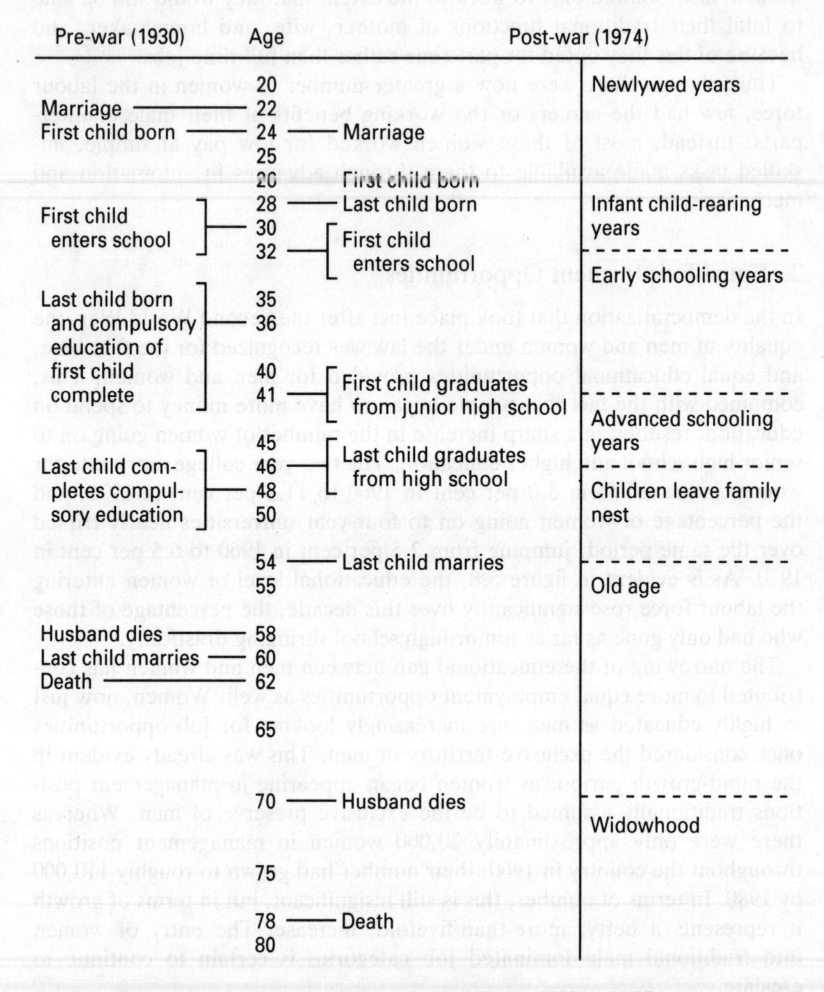

Fig. 5.9. Changes in the Japanese Woman's Life Cycle, 1930 and 1974

Note: The average number of children before the war was five, after the war around two.

Sources: Prime Minister's Office, *National Census*; Ministry of Education, *Gakkō kihon chōsa* [Basic Survey of the Schools]; Ministry of Health and Welfare, *Nihon jinkō no dōko* [Trends in Japanese Demographics] and *Jinkō dōtai tōkei* [Population Movement Statistics]; Shōrai Kōsō Kenkyūkai, ed., *Zusetsu onna no genzai to mirai* [Women, Present and Future: Illustrated] (Aki Shobo, Tokyo, 1979), adapted from p. 139.

Of basic concern to women today is how to spend the many years they have free, years their mothers and grandmothers never had, most productively. This is one of the reasons you find so many women in continuing education programmes and involved in a wide variety of hobbies and interests of all kinds including dance, swimming, tennis, and much more. There have been negative effects of this trend. Many mothers pour their excess energies into their children, leading to overprotectiveness and an excessive concern for their children's educational advancement. Other middle-aged and older housewives, finished with childrearing and lacking a truly close relationship with their company-oriented husbands, are swept by a sense of worthlessness and sink into a depression that is generating numerous physical and psychological disorders peculiar to the modern Japanese housewife.[24]

Today's Japanese woman wants to work not simply for economic reasons, but because she seeks money to finance her leisure activities or because she is looking for something worthwhile and interesting, some way of participating in society. Now that they can expect to live well into their eighties, Japanese women are incorporating work into their lifestyles, considering it as important as marriage, childbearing, and homemaking.

The rapid-growth era saw an exodus of young male workers from the farming communities to the cities, where they became salaried employees. As these men married, the phenomenon of small, nuclear families emerged. Women were expected to concentrate on the household and their children, and were idealized as the perfect helpmate whose function was to create a comfortable home so that the man, the breadwinner of the household, could concentrate on his outside work. This ideal image of women was emphasized in women's magazines.

The assumption that a woman's place was in the home was accepted without question throughout Japanese society, and companies only viewed women as a temporary labour force to be tapped before marriage. Female university graduates were considered to have very few workable years before marriage, and women who aspired to true careers found very few such opportunities.[25]

In general, women workers of this period felt little commitment to their jobs. In response to a survey of female office workers, 46 per cent responded that they would quit upon marriage, and 30 per cent upon having children. Only 5 per cent expressed any interest in continuing work even after marriage and pregnancy.[26] This lack of commitment, however, related directly to corporate attitudes toward female workers; it was the general practice among private companies during the rapid-growth era to retire their female workers when they married or got pregnant.

By the latter half of the rapid-growth period, frustration at the discrepancy in retirement practices and wage scales for men and women were causing more and more women to take their complaints to the courts. Landmark cases in which retirement practices applying exclusively to women were declared discriminatory and illegal include the December 1966 ruling against

Sumitomo Cement's policy of retiring female workers at marriage; the March 1977 ruling against Mitsui Engineering and Shipbuilding on the issue of requiring women to retire when they became pregnant; and the August 1975 ruling against the Shaboten Koen (Cactus Park) authorities' setting of different retirement ages for men and women.[27] By 1983, only 18.5 per cent of all private corporations had different retirement policies for men and women workers.

The gradual erosion of long-held distinctions between male and female workers resulted directly from the fact that more women are working for longer periods of time at higher-level jobs than ever before, and, though they are still a minority, these women are beginning to find the voice to assert their rights and demand equal treatment with men.

V. Conclusion

On 17 May 1985, the Equal Employment Opportunity Law was passed. Though this law does not do much more than urge corporations to make an effort to treat their female and male employees in exactly the same way, it does indicate a shift toward the elimination of discriminatory practices regarding employment, promotion, training, and firing practices. The new law replaces the special protective measure that used to be applied to female workers under the Labour Standards Act, but in actuality neither corporations nor their female workers are very optimistic about its application.

In a survey made just before the law went into effect,[28] a majority of corporations said it would not affect their policies and that women workers would have to show greater job commitment first. Among female workers, it was found that two-thirds did not even know about the proposed law and those who did felt that legal elimination of discrimination depended upon their employers' efforts.

The movement toward equality between the sexes in both the home and the workplace and the changing interpretation of the traditional division of labour is an international one. Equality in employment was the premise of the agreement, to which Japan was a signatory, to eliminate sexual discrimination against women at the World Conference of the United Nations Decade for Women held in Copenhagen in 1980. Japan is being compelled to follow this trend as new technological innovations and the development of the information society rapidly move it toward greater internationalization.

The long-accepted axiom, "man's place is in the workplace; women's in the home," no longer holds true, and the time has come for greater flexibility in allocating jobs to both men and women. Women should no longer have to endure lower wages and inferior working conditions.

Rising incomes, modernization of lifestyles, and the still-falling birth rate have all contributed to narrowing the educational gap between men and women and to increasing the importance of work within a woman's life cycle. The entry of married women into the workplace has led to a new consciousness that the husband and father has a role to play in the home after all. Outside the home, society is gradually coming to accept the fact that women, too, can have lifelong careers and can fulfil just as competently functions previously reserved exclusively for men.

The female labour force that emerged during Japan's years of rapid economic growth thus had an effect upon employment that extended beyond the simple increase in the number of women workers to encompass a major transformation in traditional employment practices and social mores. It is hoped that corporations will continue with this change, that the government will encourage it, and that the Japanese people as a whole will develop a heightened awareness of the need for greater equality in the labour market.

Notes

1. *Rōdō hakusho* [White Paper on Labour] (Ministry of Labour, 1984), pp. 14–16.
2. During this period, automation and other technological advances pulled Japan out of the aftermath of war, transforming the defeated nation into one with the second-largest GNP in the world. Five basic reasons are generally given to explain this phenomenon: (1) private industry's investments in equipment and facilities coincided with major advances in technological innovation; (2) industriousness, a national trait, combined with a high level of education, created a superior labour force; (3) the Japanese people's propensity to save helped to fund large-scale capital investment; (4) the government adopted growth-promoting measures, providing investment financing and encouraging exports; and (5) postwar democratization stimulated the Japanese people to greater productivity and higher consumption.
3. Ministry of Labour, "Daisanji sangyō koyō jittai chōsa" [Employment Conditions in Tertiary Industry], *Fujin rōdō no jitsujō* (1981), p. 97.
4. Figures for 1956 and 1961 from Ministry of Trade and Industry, *Industral Statistics*; for 1970 from Ministry of Labour, *Monthly Labour Statistics*.
5. Technological innovation introduced from overseas between 1949 and 1956 included, by industry, 143 innovations in electrical appliances, 148 in all other kinds of equipment, and 146 in chemicals. These three industries combined accounted for 66 per cent of all technological innovations adopted from abroad, while the spinning industry accounted for only 6 per cent, or 37 innovations. From Hoshino Yoshirō, *Gijutsu kakushin no konpon mondai* [Basic Issues of Technological Innovation] (Keisō Shobō, Tokyo, 1958), p. 247.
6. Ministry of Labour, Women and Minors Bureau, *Fujin rōdō no jitsujō* (1981), p. 11.
7. Some readily available works on the issues of eliminating sexual discrimination and promoting equal employment opportunities include: Higuchi Keiko et al., *Shokuba, hataraki tsuzukeru anata e* [The Workplace: For Career-minded Women Workers] (Chikuma Shobō, Tokyo, 1982); Takenaka Emiko, ed., *Joshi rōdō ron* [A Study of Female Labour], (Yūhikaku, Tokyo 1983); Ōba Ayako and Inoue Shigeko, eds., *Josei ga hataraku toki: Hogo to byōdō to* [When a Woman Works: Protection and Equality] (Miraisha, Tokyo, 1984).

8. Employment Promotion Project Corporation, Fujin Koyō Chōsashitsu [Women's Employment Office], *Joshi rōdōryoku no dōkō to joshi ni atarashiku hiraketa shokushu, Shōwa 38nen–Shōwa 43 nen* [Trends in the Female Labour Force and New Job Categories, 1963–1968] (1969). Basic divisions by industry are: the production division in manufacturing industry; the administrative divisions in the electricity, gas, and water supply industries; the services division in transportation and communication; the administrative division in finance and real estate; and the services division in the service industries.
9. Ibid., based on case-studies of new job categories.
10. Furukawa Sachiko, "Denki sangyō ni okeru fujin rōdō" [Female Workers in the Electrical Appliance Industry], in Ōba Ayako and Ujihara Masajirō, eds., *Fujin rōdō* [Female Labour] (Aki Shobō, Tokyo, 1969), pp. 213–240.
11. Shimazu Chitose, ed., *Gōrika to joshi rōdōsha* [Rationalization and Female Labour] (Rōdō Shunpōsha, Tokyo, 1965), pp. 10–15; and *Ruporutaaju shokuba* [Report: The Workplace] (Shin Nihon Shuppansha, Tokyo, 1971), pp. 43–71.
12. Ministry of Labour, Women and Minors Bureau, *Seimitsu kikai kigu seizōgyō no joshi rōdōsha* [Female Labour in the Precision Machinery and Tool-making Industries] (1962). Other references include the following reports made by the Bureau on female labour in the changing industries of the rapid-growth era: *Pankashi seizōgyō no joshi rōdōsha* [Female Labour in the Bread and Confectionery Industry] (1959); *Suisan shokuryōhin seizōgyō no joshi rōdōsha* [Female Labour in the Marine Food Production Industry] (1960); *Kinzoku kikai seizōgyō ni okeru fujin rōdō jittai chōsa* [Female Labour in the Metal and Machinery Industry] (1972); and *Sen'i kōgyō ni okeru fujin rōdō jittai chōsa* [Female Labour in the Textile Industry] (1973).
13. Employment Promotion Project Corporation, Fujin Koyō Chōsa Kenkyūkai [Women's Employment Research Centre], *Fujin no shokugyō bunya toshite no konpyuutaa kanrenshokushū ni kansuru chōsa kenkyū* [Study of Computer-related Occupations as Employment for Women] (1970).
14. In 1964, the keypuncher's complaints were formally recognized as a kind of occupational disease, and work standards were established by the Ministry of Labour. For details, refer to Saitō Hajime, ed., *Keikenwan shōgai to yōtsū* [Tenosynovitis and Lumbago] (Rōdō Kagaku Kenkyūsho, Tokyo, 1979).
15. Studies on the effects of office automation on women workers have only recently been carried out. They include: "Microelectronics kakumei to Nihon no josei rōdōsha" [The Micro-electronics Revolution and Women Workers in Japan], a paper presented at an international conference on New Technologies and Women by the Computer to Josei Rōdōsha o Kangaeru Kai [Study Group on Computers and Female Labour], June 1983; and *OA, Jōhōka no josei rōdōsha e no eikyō chōsa* [Study of the Effects of Office Automation and Information Processing on Women Workers] (All-Japan Federation of Electric Machine Workers' Unions, 1984).
16. Today part-time workers fulfil a wide variety of functions, many working the same hours as full-time workers. Other types of non-full-time workers include short- and long-term temporary employees, regularly hired part-time workers, contract workers, commissioned workers, etc. Recently studies have been made of employment trends for these kinds of non-full-time workers and how they are being affected by the changing industrial structure. These studies include: *Fuantei Shūgyō to shakai seisaku* [Unstable Employment and Social Policy] (Ochanomizu Shobō, Tokyo, 1980) and *Chūshō jigyōsho ni okeru hi-seiki jūgyōin no jittai chōsa*

[Study of Non-full-time Employees in Small Offices] (Tokyo Labour Institute, Tokyo, 1981).

17. Ministry of Labour, Women and Minors Bureau, *Joshi paato taimu koyō no jitsujō* [Employment of Female Part-time Workers] (1971), pp. 18–19.
18. Hashimura Hideaki, Labour Section, Hitachi Mobara Plant, "Pāto-taimā kanri no kotsu" [Managing Part-time Workers], *Rōmu jijō* (Industrial Labour Institute), no. 133 (1968).
19. Shinozuka Hideko, "Joshi paato-taima no saikin no dōkō" [Recent Trends among Female Part-time Workers], *Shokken* (Employment Research Institute), no. 32 (1980): 6.
20. Figures for 1965 are taken from Ministry of Labour, *Joshi paato-taimu koyō chōsa* (1966); figures for 1982 are taken from p. 58 of Ministry of Labour, *Fujin rōdō no jitsujō* (1983). Rapid advances have been made in studies of part-time workers over the past few years. Outstanding examples include Japan Federation of Textile Industry Workers Unions, *Pāto-taimā no jittai to ishiki chōsa* [Study of Part-time Workers and Their Attitudes] (1980), and Zenrōdōshō Rōdō Kumiai, *Joshi pāto-taimā jittai chō* [Female Part-time Workers] (1982). The former is a detailed study of typical part-time job categories and the latter is a national survey.
21. For further information on the labour unions and their dealings with part-time workers, see Ōsawa Masanori, "Izumiya ni okeru pāto soshikika no torikumi" [Organizing Part-time Workers in Izumiya], *Kikan rōdōhō*, no. 117 (1980); Nakamoto Takahisa, "Shōgyō Rōren ni okeru pāto-taimu shain no soshikika" [Organizing of Part-time Workers by the Japan Federation of Commercial Workers' Unions], *Rōmu jijō*, no. 502, (1980); and Sugai Yoshio, "Zensen Dōmei no pāto-taimā soshikika no tame no gutaisaku" [The Japan Federation of Textile Industry Workers' Unions' Proposals for Organizing Part-time Workers], *Rōmu jijō*, no. 502.
22. As reported in the evening edition of the 17 September 1984 *Nihon keizai shimbun*. The *Pāto-taimā hakusho* [White Paper on Part-time Workers] (1984), put out by the Sangyō Rōdō Chōsajo [Industrial Labour Institute], presents a comprehensive view of recent part-time labour activities.
23. In 1960 there was a total of 9,782 public and private nurseries with capacity for 689,242 children. By 1973, the number of public nurseries had been increased 1.7 times to 16,411, expanding the total nationwide capacity to 1,425,637 children, a 2.1-fold increase. However, according to Fuse Akiko ("Shufu ga shigoto o motsu koto" [When a Housewife Holds a Job], in *Gendai Nihon no shufu* [The Contemporary Japanese Housewife], NHK Books, 1980), in 1976 there were 1,680,000 households with 2,270,000 children requiring child-care facilities, yet only nursery facilities for 1,800,000 children, leaving some 470,000 infants and children with nowhere to go.
24. Saitō Shigeo, *Tsumatachi no shishūki* [Housewives' Midlife Crisis] (Kyōdō Tsūshinsha, Tokyo, 1983). According to a report in the 17 June 1983 *Asahi shimbun*, the National Health Research Institute found in a study that full-time housewives in their late forties often suffered from an extreme dislike of housework, depression, and irritability.
25. See the comments made by management and female student representatives in "Joshi gakusei no shūshoku mondai o kangaeru" [Employment Issues for Female College Graduates], *Rōmu jijō*, no. 49 (1965). In around 1961 there was considerable debate over the so-called "corruption" of the many women graduates

who were then entering the workforce in large numbers. A sound refutation of this idea, supported by facts, is given by Fujitani Atsuko and Uesugi Takami, eds., *Daisotsu josei hyakumannin jidai* [One Million Women College Graduates] (Keisō Shobō, Tokyo, 1982).

26. Prime Minister's Office, Secretariat's Public Relations Office, "Joshi jimushokuin no ishiki chōsa" [Survey on Attitudes of Female Office Workers] (1962), as reported in *Rōmu jijō*, no. 115 (1967), pp. 6–11.
27. Akamatsu Yoshiko, ed., *Kaisetsu joshi rōdō hanrei* [Judicial Precedents Relating to Female Labour] (Gakuyō Shobō, Tokyo, 1976). Women only began to win such cases after many of them had become a long-established presence in the workplace. Among these forerunners were a number of women forced by the war to remain single and to support themselves. Also see Tani Kayoko, *Onna hitori ikiru: Dokushin sabetsu no naka o ikinuku chie* [A Woman Alone: Learning to Live with the Discrimination against Single Women] (Mineruba Shobō, Tokyo, 1982).
28. The Recruit Corporation surveyed 1,538 women working in private companies within the Tokyo metropolitan area on their attitudes toward the proposed equal employment opportunity law. The Industrial Labour Institute surveyed corporations on their response to the proposed law and also made a study of the law's effects on personnel management immediately after it went into effect. The results of both surveys were presented in *Rōmu jijō*, no. 619 (1984). Further studies are certain to be made on equal employment practices. In the meantime, *Jurisuto, no. 819-Danjo koyō kintō hō* [The Jurist, no. 819: The Equal Employment Opportunity Law] (Yūhikaku, Tokyo, 1984) gives some perspective on the labour laws involved, while *Fujin rōdō ni okeru hogo to byōdō* [Protection and Equality for Female Labour] (Keibunsha, Kyoto, 1985) looks into the social ramifications.

Chapter———6

Conclusion

Masanori Nakamura

I

Female labour has always held a lower status in society than male labour. Considered auxiliary, female labour has provided the source of supplementary income for the household. For these reasons, women's wages have invariably been low and working conditions poor. Even today, there is no fundamental change in this situation.

In the October 1984 Labour Ministry's White Paper on Female Labour, the number of working housewives exceeded that of housewives with no outside employment for the first time. Of a total of 30,420,000 married women, 15,310,000, or 50.3 per cent, were employed, compared to 14,720,000 who were housewives exclusively. Thus Japan, like other industrialized countries, has entered the era of the "working housewife." One of the most marked changes was that 60 per cent of working women were 35 or older, and the trend for working women to be older, to have better educational credentials, and to work for longer periods has become well established. The number of part-time workers, whose employment status is not secure, also exceeded the 3 million mark for the first time. Women's working conditions continue to be hard, as indicated by the fact that their average wage is only 70 per cent that of men for persons in their forties. Issues involving female labour are certain to take on increasing importance in the years to come.

Although the subject of female labour has begun to attract interest in recent years, this book is one of the few that examine the issue from the viewpoint of technological innovation. Little prior work has been done in this field and I believe that through this book we have been able to add some important research. This final chapter briefly summarizes the content of each chapter.

Female labour came to play an important role in Japan's industrialization process between 1890 and 1910, the years of the industrial revolution.

In the 1900s, female labour accounted for 60 per cent of the industrial labour force, which numbered 800,000. During this period, women worked mainly in the cotton-spinning, silk-reeling, weaving, match-manufacturing, tobacco-manufacturing, straw-plaiting, figured rush mat-making, and coal-mining industries.

The introductory chapter divided female labourers in these industries into six types, and observed the impact of changes in the industrial structure and of production technology on these groups. This book is probably the first in which precise data are presented to trace how the female labour force that took shape during the industrial revolution changed through the First World War period and up to 1945. Since the industrial revolution, female labour had been typified by women working in the textile industry, but new types of female labour in heavy industry emerged during the Second World War years. These new types of female workers were relatively better educated urban dwellers, who can be considered the prototype of the young female labour that appeared in the metalworking and machine industries during the post-war rapid economic growth period. This workforce, in fact, provided the link between the pre- and post-war years.

Chapter 1 deals with female labour in the silk-reeling industry, one of the most heavily researched areas. The study included here takes into special consideration the following points. While the silk-reeling industry has been frequently examined in recent years from the viewpoint of "the social history of technology," studies have concentrated solely on technological change and paid very little attention to the labour force itself. However, the concept of labour productivity incorporates three elements: the object and the means of production, and the workers themselves, and no study that fails to examine the living labourer can be considered a "social history" of technology. Accordingly, Chapter 1 analyses the technological changes that occurred in silk filature and the impact of those changes on the labour force. We discovered that while the level of technology is a determining factor in the quality of the labour force, that dynamic could also be reversed, with the quality of the labour force playing a major role in determining the level of technology. In the silk filature industry, the introduction of mechanized reeling devices (*setchoki*) had a particularly significant impact. Reeling is a basic production process accounting for approximately 50 per cent of silk filature work. The mechanization of silk-reeling technology necessarily meant mechanizing the *setchoki*, but Japanese silk-reelers were unenthusiastic about this prospect until very late, first because they lacked adequate financial resources, and second because, even if they had had the necessary capital, they viewed the introduction of costly *setchoki* devices as a loss as long as skilled, cheap female labour was in plentiful supply. This explains the delay in spread of *setchoki* use. In 1919, the V-shaped revolving *setchoki* was invented, but it took more than a decade before it became commonplace in large filature works. This situation basically defined the status of silk filature workers. The harsh working conditions, low wages, and long working hours, as well as the requirement to live in company dormitories and the piece-rate wage system, were closely connected to a level of silk-reeling technology that depended on the manual dexterity and skill of young female workers.

Chapter 2 focuses on female labour in the coal-mining industry, which has been studied less extensively than male labour. Female labour in this industry took hold only after the industrial revolution. The coal industry had been mechanized only to the extent of introducing a conveyance system for hauling the ore. The coal-mining process itself was not mechanized but continued to rely on manual labour and rudimentary tools. Husband-and-wife work units were predominant, with the husband (*sakiyama*) digging out the coal with a pickaxe and the wife (*atoyama*) hauling the coal to the main shafts. The *naya* (stable) system came into being to supervise labour in these difficult conditions. With increasing mechanization of coal-mining in the 1920s, female labour was eliminated. Coal-mining became fully mechanized because of the need to streamline operations, as the market price of coal had dropped owing to chronic recession and imports of foreign coal were increasing. An additional factor was the introduction of protective legislation for coal-miners. Important technological advances made in coal-mining in the 1920s were the change in extraction method from the pillar method to the longwall method; the spread of blasting; and the introduction of mechanized tools and mechanized haulers. The elimination of female labour from coal-mining occurred relatively smoothly in this streamlining process, and was facilitated by policies promoting side jobs for women who had lost their mining jobs and large-scale efforts by employers to provide welfare and recreation facilities and to unify workers in regional and family groupings. But when the economy was placed on a wartime footing after 1937, labour shortages grew and women began to work in the coal mines once again. The main achievement of this chapter is that it clarifies the distribution and roles of female labour in coal-mining from the industrial revolution to the war years. In particular, it analyses female coal-mine labour for the first time in terms of age, educational level, and number of years of employment, using the 1924 *Rōdō tōkei jitchi chōsa hōkoku* [Report of a Survey on Labour Statistics] by the Cabinet Statistics Bureau, and provides, also for the first time, a comparative analysis of miners characterized as diggers and dressers.

Chapter 3, on female workers of the urban lower class, analyses the changes in the employment structure of urban lower-class female workers in major cities (mainly Tokyo and Osaka) from 1870 to the 1910s, in connection with the advance of industrialization. Research focusing on the urban lower class usually follows the approach either that all of Japanese pre-war society belonged to the lower class or that factory workers employed in large plants rose out of the urban lower class at around the time of the First World War.

Chapter 3, however, attempts to show that the urban lower class had its own dynamic—that is, its composition changed in accordance with industrialization, urbanization, and economic fluctuations, and the employment structure also changed substantially. From the 1870s and 1880s, during the so-called primitive accumulation stage of capitalism, the employment structure of the urban lower class was at its most varied. Female labour fell into the categories of physical labourer, artisan, miscellaneous labour, and industry. Reflecting the spreading industrialization during the industrial revolution, urban lower-class women engaged in physical labour were absorbed into the match-manufacturing and textile industries. At this time, female labour was low-skilled, low-paid, and overworked.

The most sweeping change in the urban lower class occurred in the boom that accompanied the First World War. In Tokyo, for example, the proportion of the population classified as *saimin* (indigents) dropped from 12.6 per cent in 1911–1912 to 3.4 per cent in 1920. The labour market became a seller's market. Lower-class urban males attained upward mobility by obtaining jobs in large-scale heavy industry plants. After this, urban lower-class male labour was made up mainly of men working in small enterprises. As household head (male) wages rose, the employment rate for wives decreased from 70 per cent in 1911–1912 to 40 per cent in 1921. The size of the urban lower class in which wives' earnings were crucial for survival decreased. But from the latter half of the 1920s, the population in the lower class began to rise again. The number of unemployed rose with the financial panic of 1927 and the depression (the Showa Panic) beginning in late 1929, and these people sank back into the lower class as workers in miscellaneous occupations. The urban lower class subsequently continued to expand and contract, following economic trends. This chapter is significant in that it examines the situation of the urban lower class from the 1870s to the 1920s. However, it was not possible to clarify sufficiently its connection with the topic of this book, technological innovation. We should not forget that one aspect of Japan's modernization was that the process of industrialization constantly renewed the urban lower class. The remark of a Brazilian researcher who heard this paper presented at a regular research meeting highlighted this. He commented that "Japan must be the only country in the world where recycling of discarded goods is a genuine occupation." I suppose that the urban lower class has developed in many different ways, depending on the country or ethnic group.

Chapter 4 examines the relation between technological innovation and female labour after the Second World War in family-run enterprises in agriculture and fisheries. Rapid economic growth after the war led to a sharp decrease in farming and fishing households. It also prompted the exodus of young and mature men to other occupations, and as a result, the proportion of female labour in such family-run enterprises increased. The most important factor behind the higher proportion of female labour was that technological innovation had progressed enough for the work involved to be taken over by women. The effects of technological innovation on labour in agriculture and fisheries, however, were not even. In agriculture, mainly rice cultivation, land improvements (such as reclamation and better irrigation), improved seed types, and mechanized equipment reduced labour requirements dramatically, which facilitated the shift in principal agricultural workers from mature males to women and older people. As a result, agriculture has become more female-labour-oriented. The situation is very different in greenhouse horticulture and coastal fisheries. In offshore fisheries especially, fishing boats have become motorized, larger, and faster, and fishing nets are now made of synthetic fibres instead of cotton. These improvements have reduced the intensity of labour as well as increasing catches. This has permitted older men to continue to fish and made limited female participation possible. In laver (*nori*) farming, new methods of propagation, new work procedures at sea, and the use of drying machines have boosted the ratio of female labour. Another salient feature of laver farming is the high ratio of households also involved in agriculture. Advances in laver cultivation techniques, in other words, have had the opposite effect of heightening dependence on agriculture. Diving, conducted by female divers collecting abalone, turban shell (*sazae*), and other shellfish, by contrast, has rejected technological innovation and centuries-old methods continue to be practised. The main feature of this chapter is its focus on patterns of female labour in family-run fishing businesses, which had previously never been examined. The labour of the female members of households engaged in fisheries is largely determined by age, composition of household, development of the labour market, and parallel businesses in which the household is engaged. This chapter elucidates female labour, with these factors in mind, in each type of fishing operation on the basis of plentiful data.

Chapter 5, the last study, focuses on the relationship between technological innovation and female labour in the rapid economic growth period after the end of the Second World War. Female employment grew rapidly during this period, particularly among middle-aged housewives. On the labour supply side, factors responsible for the quick rise in the female labour force included less time required for housework, increased spending on consumer durables, greater attention devoted to leisure and children's education, and the consequent need for supplementary income to meet such expenses, as well as changes in the women's life cycle. On the labour demand side, the shortage of male labour, and mechanization and automation as a result of technological innovation, created more job opportunities for women. The influx of female labour was especially marked in manufacturing, clerical work, services, and sales. A large proportion of labour was employed part-time, for short periods and on a non-regular employee basis. Many housewives worked as part-timers because social conditions made it difficult for them to work as regular employees. It was also more advantageous for employers to hire cheap part-timers who could be let go when they were no longer needed. Many young women took full-time jobs in the electrical and precision machinery industries, but most of the work involved simple unskilled tasks of an auxiliary nature, and only a rare few were employed in positions requiring specialized skills. In the computer industry, the work was monotonous yet demanding, causing considerable work-related fatigue, and the turnover rate was very high.

Disparities in the wages paid to men and women continue. While the rapid economic growth period presented greatly expanded work opportunities for women, the fact that it also gave rise to new types of gender discrimination should not be overlooked. In addition, changes in women's life cycles have produced a situation where women today, unlike their own mothers, have many years of vigorous life left after they finish raising their children. Finding ways of using these years productively is not simply a question concerning middle-aged and older women, but an important issue for all of Japanese society.

II

As stated in the Introduction, the aim of this study on "Technological Innovation and Female Labour" is to summarize the Japanese experience for the benefit of developing countries. However, circumstances in Asian developing countries are, needless to say, quite different from those of the Japanese experience. The urban lower class in these countries is certainly expanding, as it did in Japan during the Meiji years, but technological progress such as that experienced by Japan in its period of rapid economic growth after the Second World War is taking place simultaneously, giving rise to problems never encountered in Japan.

It is common to classify the Asian nations in groups like NIEs (newly industrializing economies), which include the Republic of Korea, Taiwan, Hong Kong, and Singapore, the ASEAN countries (including Malaysia, Thailand, Indonesia, and the Philippines) and the South Asian countries (India, Pakistan, Bangladesh, and Sri Lanka). Degree of industrialization, social structure, and ethnic traditions differ not only between these groups but also between countries in each group. And as far as female labour is concerned, conditions vary vastly. The male labour force, by contrast, follows virtually similar patterns in both the industrialized and the developing countries (fig. 6.1).

Table 6.1 is a compilation of statistics showing percentages of female labour in the 11 Asian countries for which data were available. In some of these countries, the figures themselves are not always reliable and rapid changes in the labour force composition do not necessarily reflect current conditions accurately. However, they do suggest some general features and patterns concerning female labour in Asia.

The first feature notable in these statistics is the many countries with a high birth rate and, consequently, a large population of young people. This is manifested in the index of the dependent population (obtained by dividing the population aged 14 and under and 65 or over by the population aged 15–64). In fact, there are many countries where a large number of children under the age of 15 work. The index shows similar ratios of the dependent population for Pakistan and Thailand, but in Thailand women constitute a very large proportion of the labour force supporting non-working members

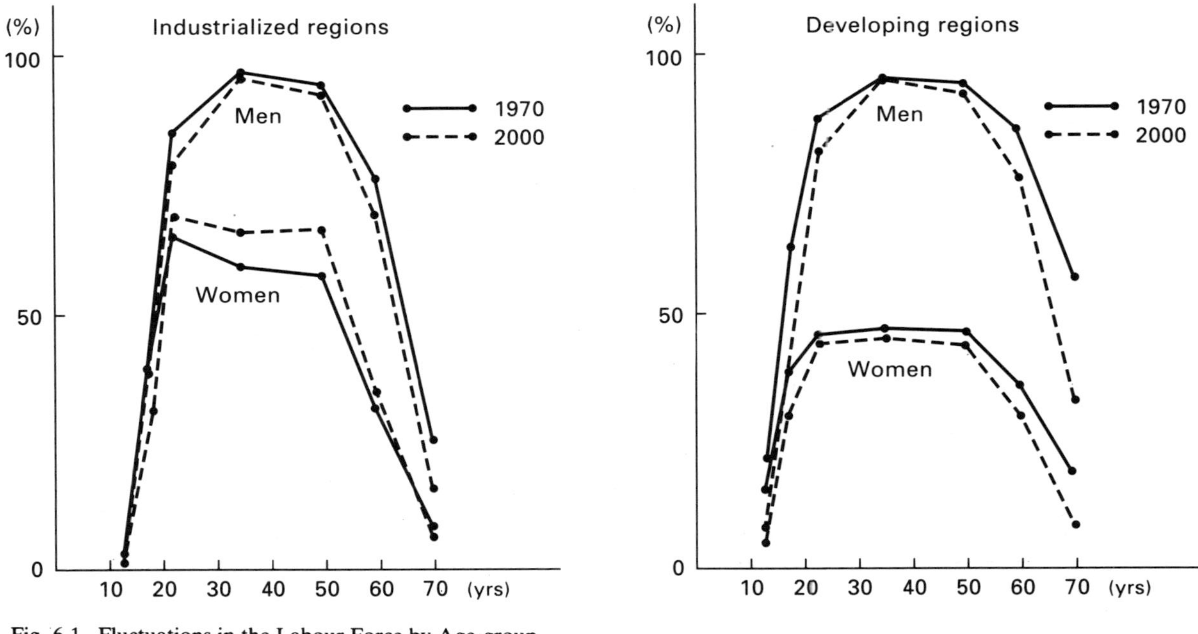

Fig. 6.1. Fluctuations in the Labour Force by Age-group
Note: Figures constitute percentages for workers aged 15 and over.
Source: Sagaza Haruo, "Ajia shokokuno jinkō mondai nyūmon" [Introduction to Population Problems in Asian Countries], *Ajiken nyūsu* [Institute of Developing Economies News], no. 38 (1983), p. 10.

Table 6.1. Composition of the Labour Force in 11 Asian Countries

Country	Dependent population (%)	Working population[a] (% of women aged 15+)	Working pop. by sector[b]			Female workers (aged 15+) (%)	Female workers by sector (%)[b]		
			Primary	Secondary[c]	Tertiary		Primary	Secondary[c]	Tertiary
Pakistan (1984)	95.2	27,740 (11.6)	50.7	18.7	26.4	11.3	36.2[d]	16.9[d]	34.9[d]
India (1981)	85.3	244,604 (26.0)	62.6	12.6	15.8	29.9	57.5	6.6	6.7
Sri Lanka (1980–81)	68.8	5,714 (28.1)	41.7	14.7	24.7	31.3	43.7	12.1	20.5
Indonesia (1980)	79.1	52,153 (33.0)	55.3	13.0	29.9	36.8	52.9	12.8	31.9
Philippines (1978, 1983)	83.3	17,362 (37.0)	49.9[e]	13.2[e]	32.7[e]	46.8	35.9	14.5	42.4
Malaysia (1980)	76.0	4,923 (33.7)	34.4[e]	24.5[e]	35.4[e]	40.1	—[d]	—[d]	—[d]
Singapore (1983)	41.3	1,208 (35.5)	1.0	36.0	62.4	45.7	0.6	36.1	62.5
Hong Kong (1983)	45.7	2,568 (36.3)	1.2	45.4	52.6	48.3	0.9	50.0	48.0

Republic of Korea (1983)	56.8	15,128 (38.5)	28.5	28.0	39.4	42.8	31.9	22.6	43.3
Japan (1983)	47.8	58,890 (39.5)	9.0	33.9	54.2	49.0	11.0	27.6	58.5
Thailand (1980)	96.6	22,728 (47.3)	70.1	10.3	18.7	76.5	73.5	7.7	18.0

a. Calculation of working population is sometimes based on the labour force formula and sometimes on the working persons formula, and varies from country to country, and sometimes for the same country in different years.
b. Some workers do not fall in any of the three sectors, so the percentages for the three sectors do not add up to 100. The same is true for the female working population by sector.
c. Mining, manufacturing, electric, gas, and water supply facilities and construction are included in the secondary sector.
d. Figures for the female working population by sector for Pakistan are for 1981; for Malaysia they are unknown.
e. Figures for total working population by sector for the Philippines are for 1983 and for Malaysia for 1979.

Source: ILO, *Yearbook of Labour Statistics* (1981, 1982, 1983, 1984).

of the population. This indicates that the pressures of the dependent population are mitigated by the high participation of women in the labour force. The dependent population ratio for China, which has implemented a one-child per couple policy, has dropped to 62.6 per cent (1982).[1] Notably, in China as well, approximately half the working population (43.7 per cent)[2] consists of women.

The second feature of the labour force in Asian countries is the large percentage employed in agriculture. With the exception of Singapore and Hong Kong, only in Japan, Taiwan, and the Republic of Korea do less than 30 per cent of the population work in primary industry. Japan's proportion of people employed in the primary sector decreased drastically from the 30 per cent mark after rapid economic growth began in the 1960s. The number of people employed in the primary sector in Asian countries is expected to change; the question remains what the impact of female labour will be. The majority of women employed in agriculture are, as was formerly the case in Japan, workers in family enterprises. As in Japan, the decrease in female agricultural workers led to a decline in the rate of female participation in the labour force. In other words, the decreasing number of farms resulted in the loss of employment for many family employees. In recent years, however, the actual number of women in the working population has been growing, especially for middle-aged and older workers in tertiary industry. Thus, over the long term, although it is affected by shifts in the industrial structure, the female labour force has increased. In China, persons working in farming, animal husbandry, forestry, or fisheries account for 73.7 per cent of all employed persons, and for 78.0 per cent of employed women.[3]

Third, there is a large discrepancy in the rate of female labour in Asian agricultural countries. In Thailand, one in two workers in agriculture is female, whereas in Pakistan the figure is one in ten. As I mentioned above, the ratio of male labour-force participation is high in both industrialized and developing countries, with very few differences among countries (except for the age-group under 19, because of the longer time spent in school in some countries). The female labour participation rate, on the other hand, varies from the 10 per cent to the 70 per cent level. In general, the female labour-force participation rate (proportion of the working population among women aged 15 and over) is high in industrialized countries, with a participation rate of 50 per cent or more in many such countries.[4] The accompanying series of graphs show the female labour participation rate in the female population for every five years of age after 15, and this clearly indicates the trends in each country. Figure 6.2 gives six graphs depicting female labour force participation rate trends by age in 11 Asian countries, ranked in ascending order. Below is a brief explanation of each graph.

I. Pakistan is a country with a very low female labour participation rate. Bangladesh has a similar profile. This type of curve is also observed in the Arab countries of Africa, the Middle East oil-exporting countries, and agricultural Central and South American countries. Restrictions imposed by the Islamic or Catholic religions inhibit women's activity in

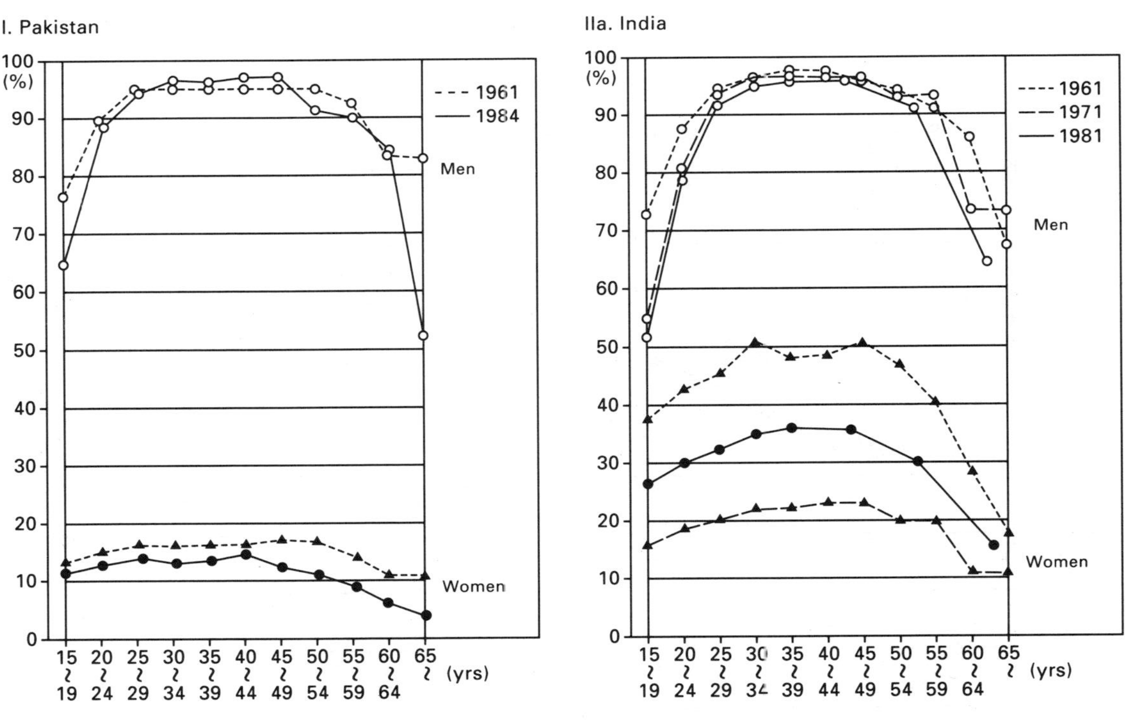

Fig. 6.2. Age-based Labour-participation Rate for 11 Asian Countries
Sources: *Sekai kakkoku jinko keizai katsudō zushū 1950–1970 nen* [Tables of Economic Activity for Individual Countries, 1950–1970], Institute of Developing Economies Statistics Series, vol. 20 (Institute of Developing Economies, 1977); ILO, *Yearbook of Labour Statistics* (1983, 1984).

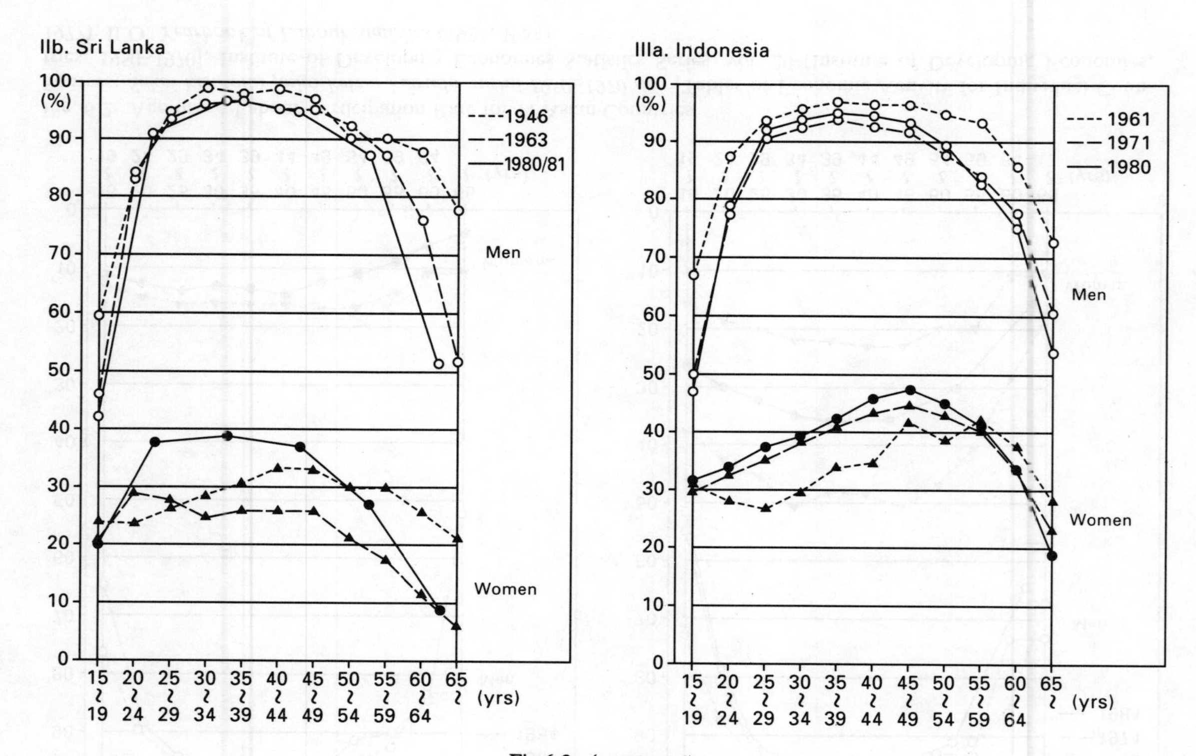

Fig 6.2. *(continued)*

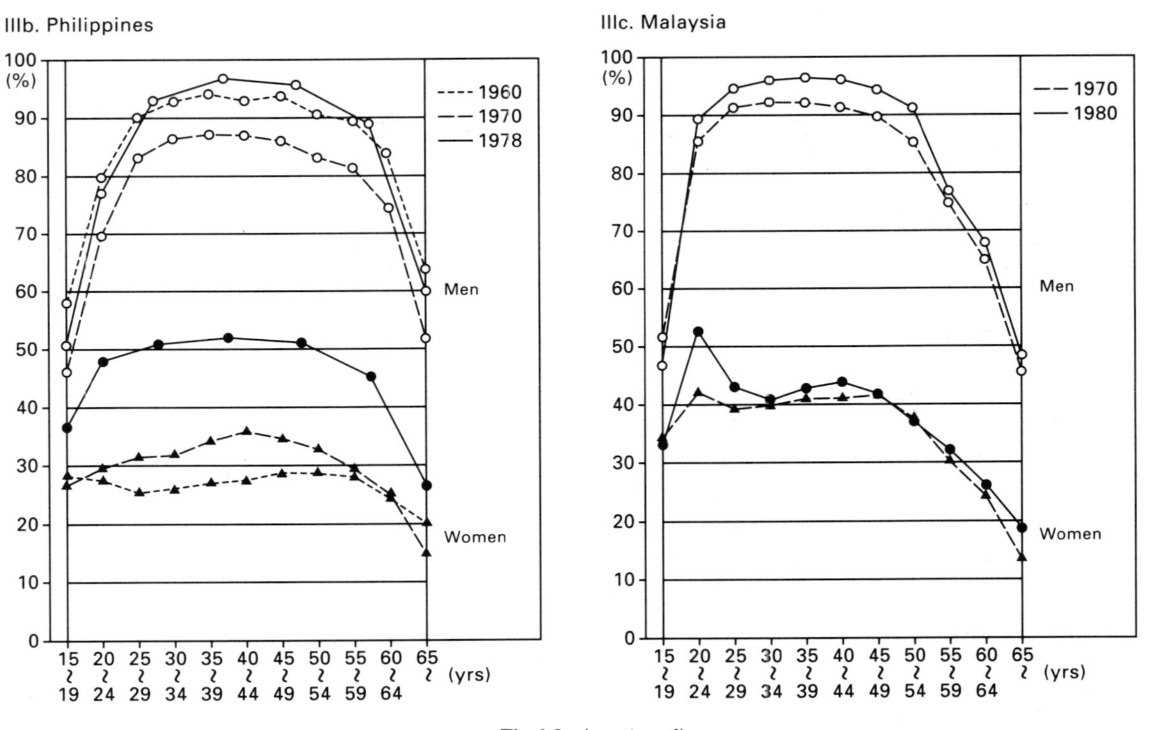

Fig 6.2. (*continued*)

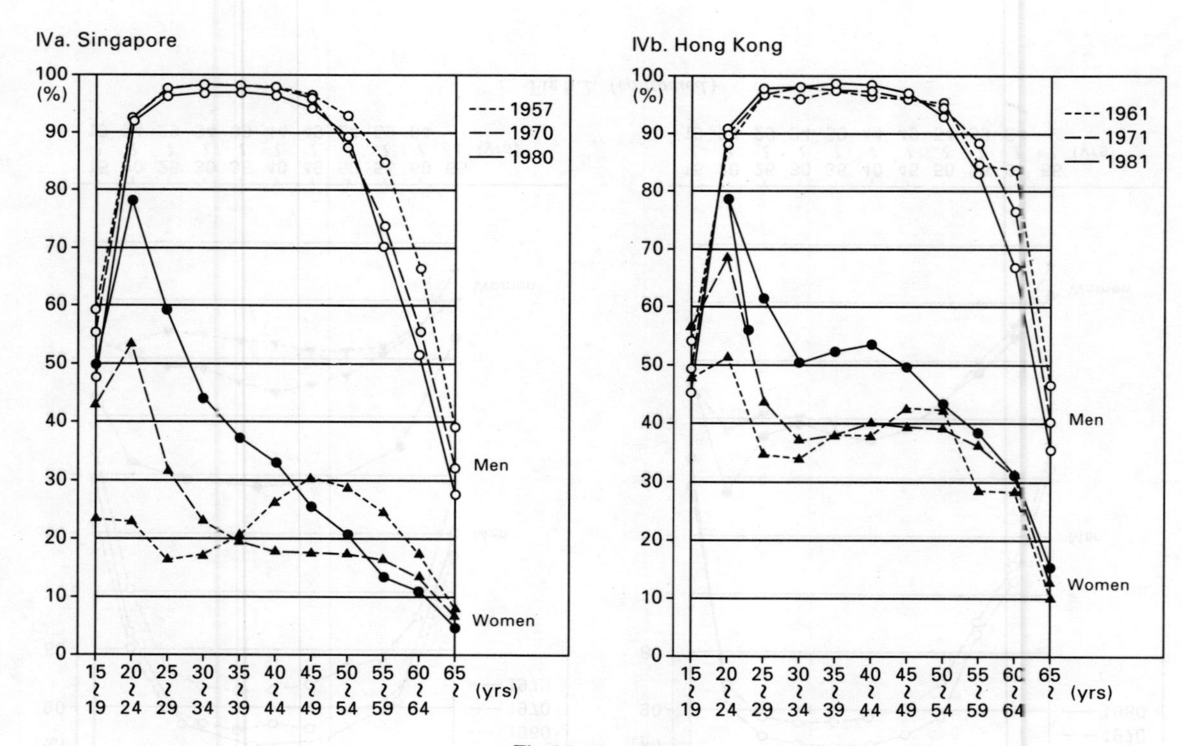

Fig 6.2. (*continued*)

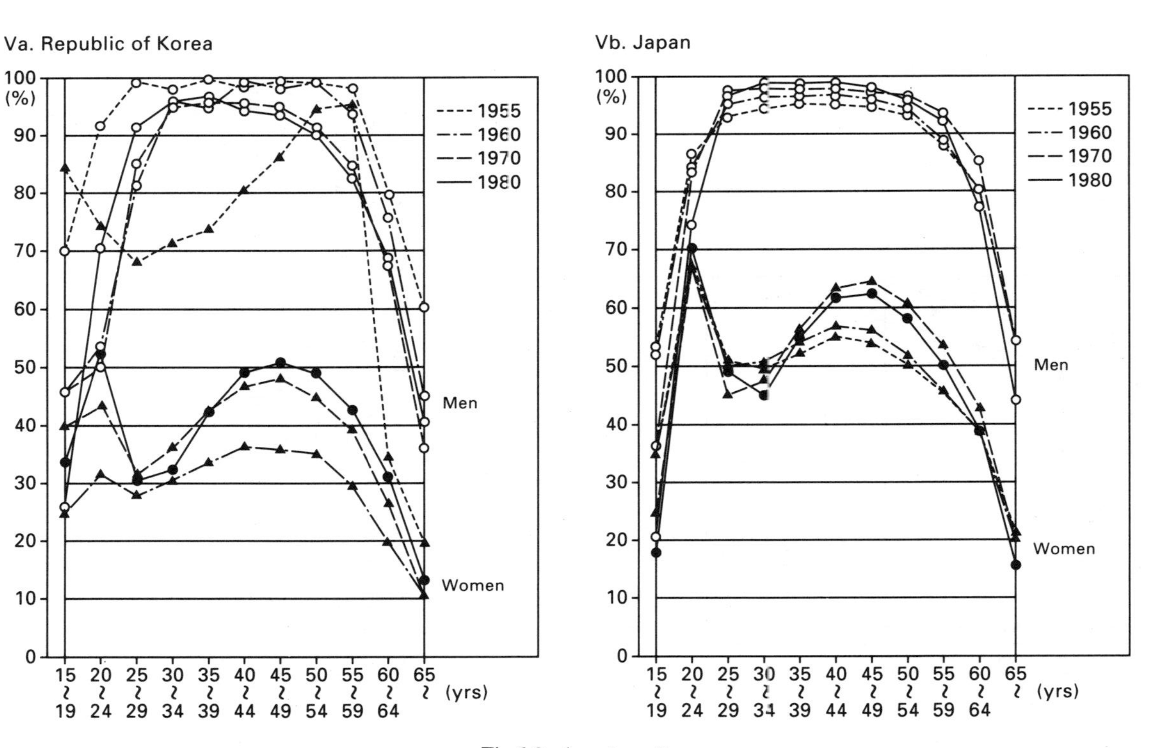

Fig 6.2. *(continued)*

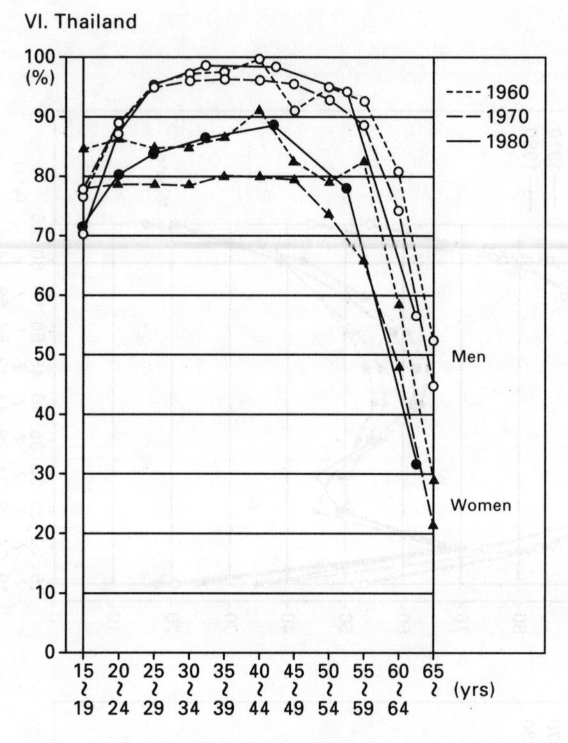

Fig 6.2. (*continued*)

society outside the family, although some Islamic or Catholic countries do have a female labour participation rate of more than 40 per cent. Female labour participation in the above-mentioned countries is likely to rise in the future, along with other changes.

II. In India and Sri Lanka, farming women account for the overwhelming majority of female labour. The gentle curve of female labour participation by age in India exhibits the typical pattern of female employment in agricultural countries. In Sri Lanka, however, the ratio of women in paid employment on plantations is twice that of women working on family farms. The pattern for Sri Lanka is also beginning to change as more young women are employed in manufacturing in export-processing zones.

III. The graph pattern shown by Indonesia, the Philippines, and Malaysia is typical of countries in transition from agriculture to manufacturing. However, the ratio of workers employed in the secondary sector, which is an indicator of industrialization, is still not very high in Indonesia and the Philippines. In Malaysia, there has been a noteworthy influx of young female labour in the home appliance and electronics industries.

IV. In Hong Kong and Singapore, which have small population and a negligible agricultural sector, the curve pattern is atypical. The young age-groups also make up the majority of the female labour, and middle-aged women do not reappear in the labour force after leaving employment for marriage. Hong Kong's pattern, however, has begun to change.

V. The graphs for the Republic of Korea and Japan exhibit a typical M-curve peaking in the young and middle-aged age-groups. This is a pattern also seen in the United Kingdom and other Western countries. Sweden had an M-curve graph in the 1970s.[5] The Sagaza paper,[6] which forecasts Asian countries' labour participation rate up to 2025 on the basis of Japanese and Korean models, concludes that India, Sri Lanka, Indonesia, the Philippines, Malaysia, and Thailand will all show M-curve patterns by that time. Only the future will tell whether the M-curve will become the typical pattern for female labour-force participation.

VI. Thailand has an extremely high female labour-participation rate among Asian countries. In that sense, its curve pattern is very similar to that of Sweden, where female labour participation (74.1 per cent in 1981[7]) is also very high. As figure 6.3 shows, except for the discrepancy generated by differences in the ratio of young people still in school, it is almost impossible to tell which curve belongs to which country. But over 70 per cent of Thai women work in agriculture, and of these 80 per cent are family workers. By contrast, 80 per cent of female labour in Sweden is employed in the tertiary sector, and over 90 per cent of all female workers are paid employees, a graphic illustration of the extremes in the conditions of female labour in these two countries. Other developing countries with high female labour-participation rates include the African nations of Rwanda, Burundi, and Tanzania. China also has a high female labour-participation rate, estimated at 70 per cent in 1982.[8]

The fourth characteristic is that female workers tend to be concentrated in the tertiary sector. In the industrialized countries, including Japan, this has become the established trend, but female labour is flowing not so much into the secondary as into the tertiary sector in some developing countries. This is indicative of the low employment absorption capacity of secondary industry in these countries. In South Asian countries, for example, the proportion of female labour is low in certain specific sectors like textiles. In other words, even the textile industry, perceived as a preserve of female labour in many countries, is a male domain in these countries. Accordingly, the condition of female workers differs not only between industrialized and developing countries, but among developing countries themselves.

In the foregoing, I divided female labour in Asian countries into six patterns according to the labour participation rate. These patterns can be said to reflect the stage of development of the labour market, employment opportunities for women (or lack thereof), and the interchangeability of

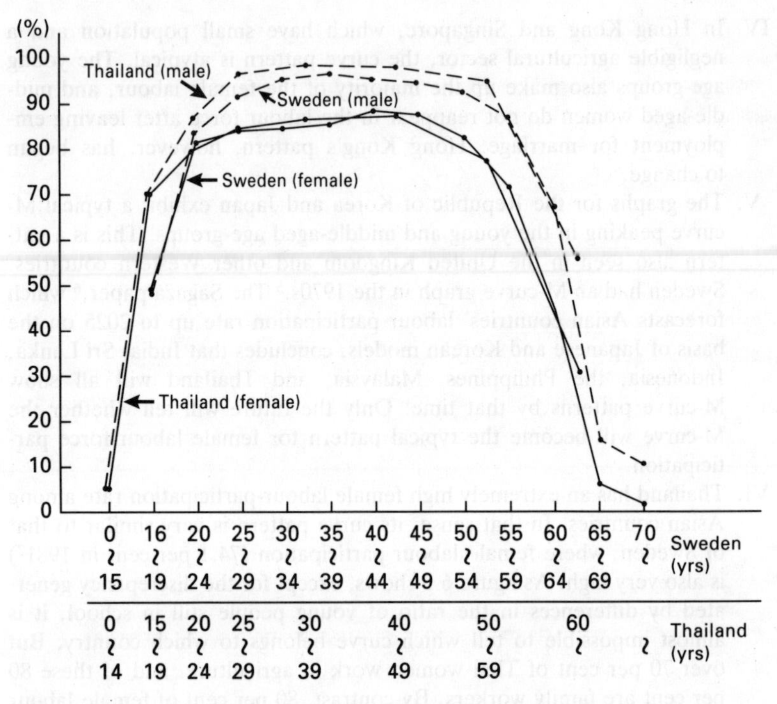

Fig. 6.3. Age-based Labour-force Participation Rate for Sweden (1982) and Thailand (1980)
Source: Graph based on ILO, *Yearbook of Labour Statistics* (1983).

male and female labour. But when factors such as rate of wage increase after hiring, pace of promotion, ratio of the population in school before employment, and type of education received are considered, these classifications could differ more. For example, in a great number of countries, differences in education dictate employment conditions for women. In India, some women with a high-school education occupy management-level positions in the national or state bureaucracies or private enterprises, and while they are paid less and advance less rapidly than their male counterparts, the gender gap is not particularly great. On the other hand, women with only meagre education are forced to work in harsh, unstable conditions as agricultural or construction labourers or as live-in domestics in the cities. This extreme divergence between the upper and lower ends of the social spectrum is also seen in Bangladesh, the Philippines, and Malaysia, and it is also possible to classify the patterns in this way. Therefore, the six classifications I set out above are merely one example. Keeping this in mind, obviously it is no simple matter to apply "the Japanese experience" in pursuing the advancement of society in the developing countries.

In Japan's case, with a few exceptions, technological innovation generally played the role of creating employment opportunities for female labour. That is, technological innovation and the increase in the proportion of female labour followed roughly parallel lines. But conditions differed greatly before and after the Second World War. In pre-war times, female labour in silk-reeling or coal mines was symbolic of the low wages earned in harsh conditions, which were endured to supplement family income. Employers were reluctant to adopt technological innovations as long as there was an abundant supply of docile, inexpensive labour. But in the post-war years, land and labour reforms were enacted and women began to receive more education, so the harsh working conditions of the "pitiful female worker" basically disappeared. In the latter half of the 1960s, when labour shortages began to appear, employment opportunities for women suddenly grew. Technological innovation and automation in the electrical and precision equipment industries made it possible for female labour to replace male labour, and these industries absorbed large quantities of young female workers. The industrial structure changes as a result of technological innovation as well, and as the larger proportion of workers has shifted steadily from primary to secondary and from secondary to tertiary industry (with over 50 per cent of the labour force in Japan in tertiary industry in the 1970s), more and more women have been employed in clerical, service, and sales jobs. The M-curve employment structure mentioned above could not have occurred without this shift in the industrial structure. Rapid economic growth fundamentally changed female labour, and female employed workers have become a major factor affecting Japan's employment structure.

More or less similar trends in female labour are found in the Republic of Korea, Taiwan, Hong Kong, Singapore, and other Asian NIEs. As industrialization progresses, these countries will probably face the problems that Japan did. In fact, some of them already have. On the other hand, in South-East and South Asian countries, where female labour constitutes over 50 per cent of the agricultural workforce, the future of female labour will be greatly affected not only by industrial progress but by land reform and changes in agricultural technology. Japan's post-war experience, particularly where land, labour, and educational reforms are concerned, bears this out.

In any case, this is the point that made us very aware, in preparing this book, that much broader-ranging interdisciplinary research will have to be undertaken before the findings of this study can be repositioned from the viewpoint of Japan as part of Asia. The United Nations University and the Institute of Developing Economies are the most appropriate organizers of such interdisciplinary research. It is my hope that the publication of this book will provide the impetus for organizing joint research that goes beyond a specialist framework.

Notes

1. *Chugoku Sōran* [General Survey of China] (Government of China, Beijing, 1984), p. 298.
2. Ibid., p. 305.
3. Calculated from *Chugoku sōran*.
4. The following paper deals in detail with female labour in industrialized countries: Kanekiyo Hiroyuki and Hayase Yasuko, "Joshi no rōdōryoku sanka to koyō, shitsugyō mondai" [Female Labour-force Participation and Issues in Hiring and Unemployment], in Minami Ryōzaburō and Mizuno Asao, eds., *Senshin kōgyōkoku no koyō to shitsugyō* [Hiring and Unemployment in Industrialized Countries] (Chikura Shobō, Tokyo, 1985).
5. Ibid., p. 99.
6. Sagaza Haruo, "Labour Force Projection for Asian Countries, 1980–2025," in Ōtomo Atsushi, Sagaza Haruo, and Hayase Yasuko, eds., *Hattentojōkoku jinkō no shōrai dōkō: kōzō to dōtai* [Structure and Dynamics of Future Population Trends in Developing Countries] (Institute of Developing Economies, 1985). The Sagaza paper makes projections based on urbanization and labour-force participation rates in Japan and the Republic of Korea. The results, as this paper also indicates, are that the M-curve pattern is likely to have a strong effect.
7. Kanekiyo and Hayase "Joshi no rōdōryoku sanka to koyō, shitsugyō mondai," p. 96.
8. Calculated from *Chūgoku tōkei nenkan* [Yearbook of Chinese Statistics] (1984), pp. 97, 102.

Contributors

Kazutoshi Kase	Professor, Institute of Social Science, University of Tokyo
Akimasa Miyake	Associate Professor, Faculty of Letters, Chiba University, Chiba
Corrado Molteni	Adjunct Professor of Comparative Economic Systems, Bocconi University, Milan
Masanori Nakamura	Professor, Faculty of Economics, Hitotsubashi University
Yutaka Nishinarita	Professor, Faculty of Economics, Hitotsubashi University
Sakiko Shioda	Professor, Takasaki City University of Economics

Index